The Mersey Tunnels: The First Eighty Years in Pictures

PETER JACKSON-LEE

AMBERLEY

First published 2017

Amberley Publishing
The Hill, Stroud,
Gloucestershire, GL5 4EP

www.amberley-books.com

ISBN 978 1 4456 6834 5 (print)
ISBN 978 1 4456 6835 2 (ebook)

British Library Cataloguing in Publication Data.
A catalogue record for this book is available from the British Library.

Typeset in 10pt on 13pt Celeste.
Typesetting by Amberley Publishing.
Printed in the UK.

Acknowledgements

Over the last four years I have met and corresponded with numerous people, most of whom were only too glad to pass on their information and experience and, without whom, the immense amount of detail in this book would have been impossible to attain. I do apologise if I have forgotten anyone, but I would like to thank the following for their time and assistance: Carl Lecky MBE, who is an accomplished author himself and gave me the inspiration to write this book, Sheena Gaskell and the staff at the Birkenhead Reference Library, as well as William Meredith, Julie Holmes and Francesca Anyon from the Wirral Archives, and Jeremy Wilkinson from the Mines and Quarries of North Wales. I would also like to thank the various staff at the Liverpool Museum who have all proven to be an invaluable source of information, and who have showed immense patience and understanding for my numerous requests for records and assistance.

Not to forget Elaine, my dearest wife, for putting up with me over the last four years whilst I researched and wrote this book in whatever spare time I had between working and other everyday commitments. I was frequently missing for large periods of time, only to be found typing or off on further research trips to the various source locations.

Lastly, and by no means least, I would like to dedicate this book to the men who have died during the construction of both tunnels. Our heartfelt thanks should go to these men in producing such marvels of engineering and for making the lives of people on both sides of the river easier.

Contents

Introduction

For many years, the only way to cross the river to either Wirral or Liverpool was to travel by ferry. By the early 1900s, the route and method of transportation for passengers and goods – including horse drawn carts – was fast becoming congested.

There are now four ways to traverse the river: by ferry, railway tunnel – opened in January 1886 by the Prince of Wales (the future King Edward VII) as the first underwater railway, and the first to be converted from steam to electricity – and through the two Mersey Tunnels; Queensway in Birkenhead and Kingsway in Wallasey. There is a fifth way to cross the river, but that would see a lengthy trip to Runcorn, which, at the turn of the century, would take the best part of the day. Even today the Runcorn Bridge is a congested bottleneck which is to receive a second tolled relief bridge. If the Queensway was being built today, a 44-foot (13 metre) diameter tunnel under a river estuary, with two branches, underwater junctions and four tunnel portals to be squeezed into city centre locations, would be an incredibly large-scale and possibly excessively expensive project.

Unlike any tunnel projects being undertaken today, the only viable prospect in 1923 was to excavate by hand. No tunnel of a comparable diameter had ever been built before. The odds were, really, stacked against its construction. The Mersey Tunnel Joint Committee was formed in 1923 and within little more than two years of feasibility and design work, they started progress on the tunnel. When it opened in 1934 it was the engineering marvel of the world.

The term 'left-footers' was applied to tunnel workers because of the noise they made as they walked to and from work across the city's cobbled streets. Digging the tunnel involved, for most workers, using their left foot to drive their shovels and spades into the ground. The result was that workers' left shoes were constantly in need of repair or replacement. To slow the wear and tear the men attached pieces of iron onto the soles of their shoes and, consequently, this caused the distinctive sound as they clattered through the streets; hence their left-footer nickname.

Work started on the tunnel on 16 December 1925 when Princess Mary set the pneumatic drills in motion to enable the first shaft to be dug on the surviving portion of the Old Georges Dock at Liverpool's Pier Head. It was from this side of the river that two pilot tunnels, one above the other, were excavated below the River Mersey. Similar excavations happened on the Birkenhead side at Morpeth Dock. Excavation was made using a combination of drilling and explosives, with a maximum of 1,700 men working on the tunnel at any one time. Seventeen of these men lost their lives and they are commemorated later in this book, as are those who lost their lives in the Wallasey Tunnel.

I have tried to stay away from the usual images you see in books and concentrated on ones that have never really been seen before or that are completely new.

Throughout the book you will see figures in brackets that represent the modern-day costs of the various project costs of 1934 and the early 1970s.

The Beginning

Once the concept of a tunnel was agreed, all the boroughs involved held numerous lengthy negotiations as to the costs and locations of such things as the entrances and exits. As expected, everyone fought their own corner in order to look after their own residents and commercial interests. As in all such cases, there was bound to be a little bit of give and take on all sides – but who was going to blink first and give the most, and who would be the eventual victor?

In March 1925, both Birkenhead and Wallasey Councils – then part of Cheshire and now part of Merseyside since the Local Government Act of 1 April 1972 – held meetings in their respective town halls and decided to support the tunnel scheme. Birkenhead had one person who opposed the scheme and Wallasey had four, but there were conditions placed upon the scheme to protect the interests of people and business within the respective boroughs.

Birkenhead Council stated they would pay only 4*d* in the £1 (£0.88 in £53), and had stipulated that the ferries were to be pooled. This was a major part of the scheme and would make the whole tunnel scheme a much more viable business option, not to mention allowing an element of future-proofing the whole scheme. Wallasey stated they would pay 6*d* in the £1 (£1.33 in £53), but the loss-making goods ferry would be taken over by the Tunnel Administration and the profitable passenger ferries would be retained by Wallasey Council. This demand was met with severe criticism as it was seen that Wallasey were laying down unfair conditions. However, Wallasey was seen to be in favour of the underground tramway, despite its possible financial loss to the footfall on the passenger ferry. They stated that they would not wish for this to be seen as an ultimatum, more a case of the borough looking after its residents and business interests.

In the days that followed and to the surprise of everybody, Sir Archibald Salvidge proposed the elimination of both the Wallasey arm and the tramways, and consequently of Wallasey itself. This action was criticised but in light of events it was probably going to be the inevitable conclusion to the events that were taking place and the demands that were being made.

Bootle Council decided to co-operate and to contribute a 2*d* in the £1 rate (£0.44 in £53). The decision of the City Council meant that responsibility for the revised scheme was to be shared by Liverpool, Birkenhead and Bootle. On that basis, a Joint Committee to promote a Parliamentary Bill was set up, with Sir Archibald Salvidge as chairman, Alderman R. J. Russell (Birkenhead) as deputy chairman, and Mr George Etherton, who had been appointed Clerk to the Lancashire County Council, as legal adviser.

The following cartoon images were published in a supplement of the *Birkenhead Advertiser* for the opening of the Queensway Tunnel in 1934:

Birkenhead Advertiser supplement image, 14 June 1934.

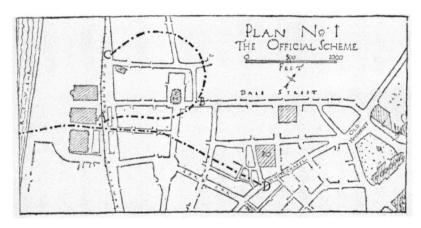

Tunnel plan showing the abandoned Whitechapel Tunnel (D). (Wirral Archives)

The acquisition of many sites included the original Carnegie Free Library, Ellis and Powell's timber yard – the replacement yard is now the Expressway Business Park at the end of the A41 – various office buildings, many shops and licenced premises, and a large steel and iron business in Cleveland Street, which was reinstated on a new site. Also included was a Rutherford's shipbuilding works in Bridge Street, founded in 1853.

Carnegie Free Library was chosen as the site for the new tunnel and was replaced by the Birkenhead Central Library. The Library was opened on the same day as the Queensway Tunnel by King George V, and is still in use today.

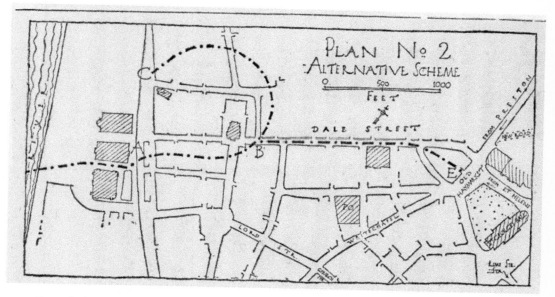

Tunnel plan showing the existing tunnel layout. (Wirral Archives)

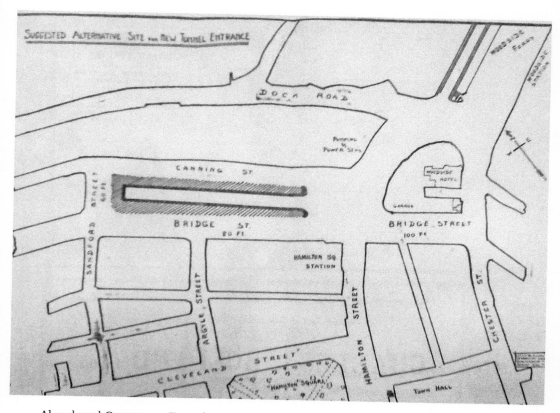

Abandoned Queensway Tunnel entrance proposal, as required by the Chambers of Commerce. (Wirral Archives)

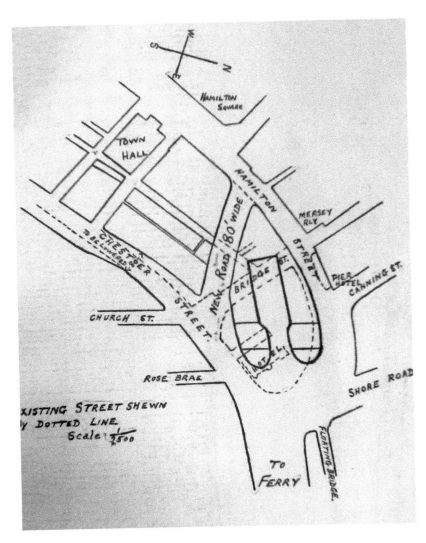

Abandoned Queensway Woodside entrance, more recently occupied by Woodside Hotel prior to its demolition following a fire in September 2008. (Wirral Archives)

Tunnel Construction

To allow the initial processes of the tunnel to be started, two shafts – one on each side of the river – were constructed up to the level of the river. Upon reaching this level, a process of silicatisation – impregnated silicate to harden a surface and reduce sensitivity to frost – and cementation treatments were added to the shaft walls. This allowed the shafts to be sunk below the river and allow the tunnel process to begin. The depths of this additional section of tunnel were 195 feet (59.43 metres) on the Birkenhead side and 100 feet (30.48 metres) on the Liverpool side.

The shafts were set on the tunnel line, and the first operations included the driving of adits – horizontal passages leading into a mine for the purposes of access or drainage – to connect with the tunnel line. The cementation of the shafts was supplemented by horizontal cementation to enable the ground, pilot holes, and connecting adits to be secure.

A small tunnel approximately 12 x 15 feet (3.65 x 4.57 metres) was started and placed on a specific heading along the correct tunnel line to initiate the pilot tunnels.

As the work proceeded, it was found that the inflow of water was not sufficient to warrant the continuation of the cementation treatment, and it was, therefore, stopped. At the time when the process was discontinued, the following areas had been covered by the process: in Liverpool, 850 feet (259.08 metres) of rock from the shaft towards the river, or 160 feet (48.77 metres) beyond the quay wall, and 450 feet (137.16 metres) from the shaft landward under Liverpool. On the Birkenhead side, the treatment had reached 1,170 feet (356.61 metres) from the shaft towards the river or 410 feet (124.97 metres) beyond the quay wall.

Old Bakery Camberdown and Hamilton Street *c.* 1929. (Birkenhead Library)

Library in Market Place South being demolished *c.* 1929. (Birkenhead Library)

Market Street / Hamilton
Street demolition 1930s.
(Birkenhead Library)

Cross Street Demolition
c. 1929. (Birkenhead
Library)

Old Feathers Hotel
in ruins. (Birkenhead
Library)

Rendell Street coal yard *c.* 1927, which was to become the now-closed Birkenhead Dock exit of the tunnel. (Author / Wirral Archives)

The Carnegie Central Library opened in 1909 and was subsequently demolished in 1929 to make way for the Mersey Tunnel (Queensway). (Birkenhead Library)

Carnegie Library at St Anne's today. This is to show the style and size in comparison to the demolished Birkenhead Library. (Author)

King Edward VII Clock (Grade 2) – At Central Station, now moved to the middle of the Central Station Roundabout due to the Queensway Tunnel. (Birkenhead Library)

King Edward VII Clock (Grade 2) – today on the Central Station roundabout and Flyover access to the Queensway tunnel from Borough Road. Birkenhead Central Station can be seen in the background. (Author)

It seems amazing that two gangs of excavators working towards each other underground, through 5,000 feet of rock, can meet at a pre-determined point below ground, but the engineer would be amazed if the results were not as planned. Given the accuracy of survey work even at this early stage, the meeting of the tunnels at the pre-determined point and at the pre-determined depth must follow as surely as night succeeds day. That the Liverpool and Birkenhead workings met with such tiny variations from the plan is merely proof of the ability and the painstaking care that engineers and contractors alike devoted to their task.

The cast-iron lining of the tunnel is composed of over 100,000 panels and weighs in excess of 80,000 tonnes. They were manufactured by the Stanton Iron Works in Ilkeston, Derbyshire, which is now closed. Stanton Iron Works' last cast took place on 24 May 2007. The tunnel segments were bolted together and the panels varied in size. A large proportion of the panels measured 6 foot in length and 2 feet or 2 feet and 6 inches in width, with the flanges being 13.5 inches in depth.

The order for the main segments was given to Stanton Iron Works on 8 July 1928 and casting was commenced on 15 July 1928, with output reaching 1,000 tonnes per week within a month. This was later to increase to 1,250 tonnes per week as the contract progressed. Because the segments were moulded in green sand, and due to their unusually large size, a considerable amount of gas was generated during the pouring of each segment.

To enable this gas to be expelled, venting was completed by hand using moulding boxes. The boxes were handled by overhead electric telpher cranes and pneumatic hoists. During the course of the contract no fewer than 248 types of cast-iron patterns were used.

The grooves for the caulking were machined out, as this was far easier than trying to cast them within the individual segment. The grooves were 1.25 inches wide (318 mm) by 0.125 inches deep (3 mm). Before any segment was dispatched to the tunnel all segments were checked, including the flange, for absolute accuracy.

At certain points the tunnel curves, and inclines had to have tapered rings formed by special segments. The segments were of an exceptional size, so it was essential that the greatest care and accuracy should be used in their production. They were required to be cast in special moulds, and the utmost care was taken to ensure that the molten metal was of the correct composition and temperature. The time allowed for the moulding was carefully calculated to obviate any imperfection in the finished casting. The panels were then machined to render them interchangeable and also to ensure the tightness of the fit when they were bolted into their final position. The importance of this had been stressed by the engineers in that a difference of one hundredth of an inch in the flange would be fatal in its final quality check.

The nuts and bolts that were required for the tunnel totalled over three-quarters of a million and weighed nearly 3,000 tonnes. In addition, washers for each bolt added another 130 tonnes to the overall weight of the materials used in its construction. The bituminous grommets that were used to seal the bolt holes totalled over three million; one-third of these being lead-faced. The seal thus secured is said to be capable of resisting a water pressure of 200 pounds to the square inch without the slightest danger of leakage.

When the lining of the tunnel had been completed, and the engineers were happy that the linings had been rendered practically impervious to water, the whole of the interior of the tunnel would be covered with a layer of 270,000 tonnes of the very finest cement. This additional layer would not only smooth out the inner surface but it would also further ensure the water-tightness of the lining.

Fortunately, Liverpool was well situated not only to receive goods, but geographically close to a good supply of materials, such as the quarries of the Penmaenmawr & Welsh Granite Company, for example, which are situated nearby. The Rhosesmor Sand & Gravel Company also supplied crushed, screened and washed gravel to McAlpine, both for the tunnel itself and for the ventilation stations that they had constructed in reinforced concrete.

Another source of supply was found in the river itself. W. M. Cooper & Sons Ltd specialised in dredging sand and gravel from the bed of the Mersey that, after suitable handling and treatment, forms an ideal base. Indeed, its quality was such that it was the only sand used by the contractors. The Gresford Sand and Gravel Company Ltd supplied their Gresford gravel, in a variety of grades.

Throughout the tunnel there was a humidity of 80 to 100 per cent. Quite apart from these difficulties and conditions, the finished surface should not be subject to unsightly crazing, so the chosen material had to withstand a high acid condition produced by the fumes from motor and steam vehicles dissolved in the salt and humid atmosphere.

The surface must be easily washed down for such things as grease, soot and other tenacious deposits, which would be considerable. A great many finishes of cement glaze, synthetic and paint types were tested, but the chosen material was 'Marplax' marble finish. Marplax would not craze or deteriorate under fair wear and usage, and it incorporated a special plaster base that was not merely a surface skin. The finishes to the periphery of the tunnel were carried out in a pale oatmeal colour, which harmonises with the dado.

During the main underwater tunnel construction, only the top half of the circle was excavated at the beginning, and it was not until the cast-iron lining of that section had been completed that the excavation of the lower half was begun. Once the chambers of the tunnel were wide enough, erectors were positioned to assist in the placing of the cast-iron segments. To allow a good work rate and reduce the risk of a collapse, the segments were placed as soon as the space would allow. In difficult ground the rock was removed with more care and in lesser quantity than before; this was to allow the maximum removal of material whilst allowing the tunnellers to maintain a safe working area.

The engineers at the outset realised that the headings for the pilot tunnels were to begin the point's remote to the Old Haymarket and Chester Street on their respective bearings. The shafts at Georges Dock and Morpeth Dock were to be constructed on a different line to the tunnel. The branch tunnels, unlike the main tunnel, were to be of a semi-circular shape and lined in cast-iron. In the case of the under-river section, the excavation below the road deck was only capable of a shallow invert and was constructed in concrete in the land section of the tunnel. The actual tunnel is 780 feet (238 metres) beyond the quay wall of St Georges Dock Parade.

As the tunnel neared the surface it had to take into account the stress of the building and structures above. The engineers had to make any necessary adjustments in consideration of the structure's safety. Added to this would be the labyrinth of utilities underneath the surface of the roads and pavements.

The first and only serious engineering set-back to the process occurred on 29 October 1930. A portion of Dale Street close to the then-police station collapsed. This was believed to be due to the line of the tunnel crossing the old fortifications of Cromwellian days. The collapse caused some disruption to street traffic for a considerable time, but did not seriously delay operations underground.

As the tunnel proceeded underneath Dale Street in Liverpool, it was initially intended to excavate without closing the street to traffic. Today this road is an exceptionally busy road through the city. The proposed method was for deep trenches beneath the pavements to the bottom level of the tunnel. The excavations would be supported, as would the roadway along with the utility services, by using large timber sections. Upon a detailed examination of the proposals, it was decided not to adopt this method of construction.

The tunnel rises at 1:300 for 167 feet (509 metres) from Birkenhead Quay wall to a gradient of 1:30.

Both sides of the river had adequate locations to store and dispose of the materials removed from the tunnel and its chambers. Unfortunately, these were not in the immediate vicinity of the tunnel, but were within a reasonable travel distance. The magnitude of this task would be appreciated when it is realised that the total amount of rock and earth removed during the course of excavation amounted to no less than 1.2 million tonnes. The rate of excavation was more than half a tonne of rock for every minute between June 1926 and August 1931.

Statistics for Birkenhead (Queensway) Tunnel

Length of tunnel	–	2.13 miles
Total length of road surface	–	2.87 miles
Total area of road surface	–	11 acres
Width of main tunnel between kerbs	–	36 feet (10.97 metres)
Width of branch tunnel between kerbs	–	19 feet (5.8 metres)
Internal diameter of main tunnel	–	44 feet (13.4 metres)
External diameter of tunnel	–	46 feet 3 inches (14.1 metres)
Cross-section area of main tunnel	–	1,680 feet (512 metres)
Lowest part of tunnel at High Water	–	170 feet (51.8 metres)
Average cover, rock, and clay underwater	–	33 feet (10.0 metres)
Weight of rock, gravel and clay excavated	–	1.2 million tonnes
Weight of excavated soil	–	560,000 lbs
Maximum amount of water from workings	–	4,300 gallons a minute

During the time of the tunnel construction, there was also a large land reclamation programme at Dingle, and the Liverpool Corporation were constructing a new river wall and embankment at Otterspool. Both sites were not only fortunate in their planning at the time of the tunnel, but were also capable of taking all the waste.

The main company selected to transport the waste material from the tunnel was Wellington Haulage who used thirty vehicles to remove over half-a-tonne of rocks and other material. The fleet of wagons were accompanied by a motorcycle to assist in the operations. This motorcycle would travel between the tunnel and dump site and check the route was clear and free from obstructions. This would allow the contractor to maximise the time and mileage to full effect.

During the night time it was only Dingle that would be used, as the conditions at Otterspool were deemed too dangerous. However, the Dingle site was only to be used with great care and the area would be illuminated with acetylene flares.

On the Wirral side, the disposal of the tunnel waste was less of a problem. Lever Brothers Ltd of Port Sunlight owned the Storeton Estate, and within this was a large quarry approximately 500 yards in length, and with depths of between 25 and 80 feet, so it was more than capable of taking most of the tunnel waste. The contractor chosen for the removal and disposal of the tunnel material was G. & W. Dodd, who entered into a contract to utilise the disused quarry for the disposal of the tunnel waste. One of the reasons this quarry was used, apart from its vast size, was its proximity to the local road network and ease of access to the tunnel.

To move the 453,000 tonnes of material, Dodd used twenty-five lorries and four steam-wagons along the three-mile route, night and day, to complete the task. Loading was completed at night with the use of electric floodlights and tipping by the use of acetylene flares similar to at the Liverpool side of operations. The quarry was also used by McAlpine & Sons, who used a fleet of six wagons for a period, with the Birkenhead Corporation also using tunnel spoil.

The walls of the main under-river tunnel carry the roadway 18 inches below the horizontal diameter of the tunnel. They are 12 inches thick, and there is a distance of

21 feet between walls, which gives a clear and uninterrupted span. The central space below the roadway and between the shafts was to be allocated to buses and trams initially, but this was later classed as future vehicle expansion. The curved passages between the walls and the sides of the tunnel were to form the vast ventilation requirements for the tunnel and were utilised as air-ducts.

In the sections of the tunnel directly underneath the land, the construction below the roadway level consisted of a shallow invert, which naturally reduced the height of the required supports. Instead of a continuous wall, a series of columns were erected at intervals of 7 feet. This allowed the space to be fully utilised as part of the fresh-air ducts.

Stanton Iron Works were chosen as the preferred contractor for the supply and fitting of the new roadway's finish. The area to be covered was in the region of half a million sections of the new proposed cast-iron surface. The surface was to be finished with ribbed tinder slices and a studded upper surface. The cast-iron segments were attached to the new concrete road by the use of bitumen type material which was also supplied by Stanton's. The bitumen was poured onto the concrete in small sections and the cast-iron segments placed on top. Once the bitumen had cooled from its approximate temperature of double the boiling point, the cast-iron sections were effectively stuck to the concrete road.

At one point in the tunnel where it passes over the Mersey Railway at Birkenhead, rubber paving blocks had been substituted for cast-iron to lessen vibration. There, 2,000 square yards of 'Gaisman' patent rubber paving blocks were laid, and were attached to the concrete roadway by a cast-iron base. These rubber blocks are black, and the process by which they are manufactured enables them to resist any cutting or tearing action which might be caused through the occurrence of a breakdown.

The tunnel roadway is first divided into two halves, one half for traffic passing from Liverpool to Birkenhead, and the other by traffic passing from Birkenhead to Liverpool. Each half of the roadway is again divided in two by the rubber blocks to separate the lane in which the slow-moving vehicles will travel from the lane to be used by fast traffic. The blocks in the central line of the roadway were substantial and were raised above the setts in a way that would cause inconvenience to any driver who may inadvertently go over

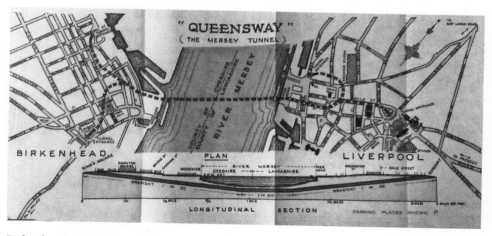

Birkenhead route selection. (Wirral Archives)

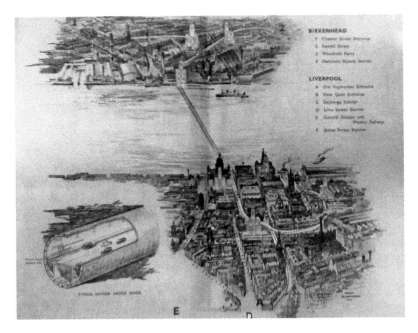

Illustration of the tunnels and its typical section. (Wirral Archives)

the line. This can be seen today on the modern-day rumble strips used on motorways and dual carriageways.

The original toll booths on the then four entrances were protected from the vehicles by rubber-covered concrete kerb units. For the surfacing of the outer gradients leading from the public highways to the mouth of the tunnel, granite setts were used. The chosen type was Bonawe granite, which was quarried and trimmed on the shores of Loch Etive in Argyle and Bute on the north-west coast of Scotland by J. & A. Gardner and Co. of Glasgow. This granite is not only of the most durable quality when subjected to heavy wear, but its surface is such that it is practically impossible to skid on. In the region of a thousand tonnes of setts were required for the works.

Birkenhead side:

1 Chester Street Entrance
2 Rendel Street Entrance
3 Woodside Ferry
4 Hamilton Square Station

Liverpool Side:

A Old Haymarket
B New Quay Entrance
C Exchange Square Station
D Lime Street Station
E Central Station
F James Street Station

Above: Starting St George's shaft.
(Birkenhead Library)

Right: Axe used by Archibald Salvidge. This
and many other items are viewable in the
Merseytravel St George's Dock building
whilst participating in the tunnel tours.
(Author)

Tunnel breakthrough –
Sir Archibald and Liverpool
and Birkenhead Mayors.
(Birkenhead Library)

St George's shaft cementation.
(Birkenhead Library)

Excavation of preliminary tunnel
prior to pilot tunnel. (Birkenhead
Library)

Primary Tunnel (same as London Tube) and Main Tunnel. (Birkenhead Library)

Cast-iron lining of preliminary tunnel. (Birkenhead Library)

Rock face and cast-iron junction (LM). Note cast-iron in top-left of picture. (Birkenhead Library)

Excavation and erection of steel ribs in one of the 'break-ups'. The tunnel was divided into sections called 'break-ups', and as the upper pilot heading was enlarged to the full size tunnel each section was lined with steel ribs. (Birkenhead Library)

General tunnel construction. (Liverpool Museum)

Tunnel section being machined. (Birkenhead Library)

Erecting cast-iron segments.
(Birkenhead Library)

Section of tunnel showing
working conditions.
(Liverpool Museum)

Cast-iron lining. (Liverpool
Museum)

Concrete filling cast-iron segments. (Liverpool Museum)

Bottom section cast-iron lining. (Birkenhead Library)

200-tonne shield (Old Haymarket). (Birkenhead Library)

Reinforcing shield under Dale Street, Liverpool. The shield protected valuable properties in the streets above from subsidence and was propelled by hydraulic rams. (Birkenhead Library)

Cement gun rendering. (Birkenhead Library)

Tunnels and dock exits under construction. (Liverpool Museum)

Completed tunnel junction. (Liverpool Museum)

Completed roadway and cast-iron lining. (Birkenhead Library)

Birkenhead entrance cover beams. (Birkenhead Library)

Rendel Street construction.
(Liverpool Museum)

Rendel Street entrance under
construction. (Birkenhead
Library)

Construction of roadway
side support walls. Note the
reinforcing bars for the full
height of the roadway supports.
(Liverpool Museum)

Road deck and supporting
walls. (Liverpool Museum)

Road deck shuttering.
(Liverpool Museum)

Road deck steel awaiting
concrete. (Liverpool Museum)

Road deck land section.
(Liverpool Museum)

Old Haymarket general
construction. (Wirral Archives)

Queensway Tunnel cable laying,
June 1933. (Liverpool Museum)

Construction of Queensway entrances. (Wirral Archives)

Side walkway and vents. (Birkenhead Library)

Applying the roadway finishes to Birkenhead entrance roadway. (Birkenhead Library)

Drainage Services and Ventilation

By its very nature a tunnel will never be leak-proof, nor will it ever be waterproof. In the case of the Queensway Tunnel, there are four large openings – one of which is now disused – which allow the weather to easily enter the complex. As the tunnel has to have a constant downwards gradient at both ends to enable traffic to drive from street-level to below the river, these openings, coupled with natural gradients and, in the case that there were no drainage solutions, would cause a large body of water collect at the invert – the lowest part of the tunnel. This would get bigger over time as there would be no sun to evaporate it.

In the event of a fire occurring in the tunnel, the water used by the emergency services must have an outlet, and the surplus water would require eventually to be discharged into the river. The drainage channels that have been constructed are, therefore, connected to sumps. One of these has been placed at each of the four entrances to the tunnel. There are two others situated in the vicinity of the Georges Dock and Morpeth Dock shafts, whilst the seventh has been constructed at the lowest point of the tunnel beneath the river. The last is the largest, and it is situated at a level below the roadway of the tunnel.

The control of all the pumps is automatic, so when the water in the sump reaches a particular level the pump will commence with pumping operations. Pumps can be turned off at any time by the control station as the level in the sumps is constantly under observation. If the need arises, any number of pumps can be activated to necessitate the removal of water up to the higher levels and into the river.

The case for both sides of the river to consider in the advance tunnel works was to avoid the dislocation of the public services. This would include public sewers, gas and water mains, as well as electrical, telephone and telegraph cables. It was necessary to involve the relevant companies to reroute the services before the tunnel workings could be carried out. This was no easy task, especially in connection with the sewerage system.

As with Liverpool, Birkenhead would see its fair share of road works carried out by the Borough Engineers Department. Similar to Liverpool, Birkenhead entrances were to see pavement and road levels changed to allow the tunnel entrances to be constructed.

There was also the consideration of the trams and the nearby train lines and junctions for Cammell Laird ship builders.

It was in the autumn of 1930 that the engineers and the committee were alarmed by a report from America that a number of people had been 'gassed', though not fatally, by carbon monoxide fumes in an inadequately ventilated land tunnel at Pittsburgh. This was quite shocking news for the committee, and the Pittsburgh Tunnel was shorter than the current Queensway Tunnel so a solution had to be found quickly.

Experiments were carried out over a period of several weeks and attended by both Dr Haldane and Professor Douglas Hay, as well as by the representatives of Sturtevant Engineering Co. Ltd and Walker Brothers (Wigan) Ltd. When the results were finally analysed, the engineers recommended the Mersey Tunnel Joint Committee to adopt the Upward Semi-Transverse system of ventilation. This abolished the necessity for the construction of an exhaust duct, which would have had to have been built below the roof of the tunnel, and so considerably reduced the capital cost of the ventilation system.

Once the final decision had been made on the ventilation system, work on its design and installation could begin. This would also allow the mechanical and electrical engineers to complete the designs for their respective installations. Once this was complete, or at least basic design conceptions had been completed, it would allow Mr Rowse to start his ventilation shaft designs; the sites of which had already been chosen and the land acquired.

Mr Herbert B. Rowse F.R.I.B.A trained at the School of Architecture at the University of Liverpool. Few people will ever realise the magnitude of the problems he had to solve. There were no precedents to guide him in the simple, masterly and impressive building of the large shafts, which were to be prominent features in the skylines of both Liverpool and Birkenhead.

There are six ventilation shaft building locations, which would be up to 210 feet high (64 metres) and would discharge the tunnel air at least 50 feet (15.24 metres) away from the intake. Mr Rowse had several basic questions to tackle, such as the size of the fans and type of machinery to be used, how the tunnel air was to be discharged without causing any annoyance to neighbouring buildings, how the machinery was to be prevented from disturbing the neighbourhood with noise and vibration, and, lastly, what the character of the external design was to be. All of these had to be answered and the buildings designed in a short space of time, so as to allow construction and completion by the time the tunnel was opened.

Steel suppliers for all six ventilation buildings – Redpath, Brown & Co. Ltd of Manchester

Few people who watched the steel framework rise into the sky could conceive the immensity of the task which was being undertaken. When the plans had been finally passed, it became necessary to construct a model showing every piece of steel that was to form part of the structure. Then the steel had to be fabricated to the required dimensions. Before it left the works, each section was marked to correspond with the working drawings supplied to the erecting staff. The speed represented a triumph of organising ability comparable only with the efficiency with which each building was completed.

Construction of the ventilation buildings

W. Moss & Sons Ltd built the North John Street station.
Henry Boot & Sons Ltd, of Sheffield, the New Quay and Taylor Street stations.
John Mowlem & Co. Ltd, of London, the Georges Dock station.
Sir Robert McAlpine & Sons, of London, the Sidney Street and Woodside stations.

Electrical installation

Higgins & Griffiths Ltd, of London, undertook Georges Dock.
John Hunter & Co. Ltd, of Liverpool, North John Street and Taylor Street.
Electric Power Installers Ltd, of Liverpool, took New Quay and Sidney Street.

Plumbing Works

R. W. Houghton Ltd, of Liverpool, were entrusted with contracts for the plumbing work required at Georges Dock, North John Street, New Quay, Sidney Street and Taylor Street ventilating stations.

Door Suppliers – The Birmingham Guild, Ltd

The special double doors to the fan chambers in all the ventilating stations were designed and fitted as an additional precaution against the transmission of sound and vibration. The outer doors were of hollow steel framing, whilst the inner doors took the form of hollow wood shutters, all being firmly clamped against vibration.

Finishes and Final Coatings – James B. Robertson & Co. Ltd, London

The final Marplax finishes were selected for the final coating of the tunnel interior and were equally successful in obtaining a contract for the supply of the final finishing surface of the walls of the fan chambers and switchgear rooms. Within the switchgear rooms, exposed concrete surfaces were treated with 'Stipplecrete' – a glazed, stippled, impervious finish.

Quantity Surveyors for all six ventilation buildings – W. M. Law & Son, Exchange Street East, Liverpool

Mr W. H. Law, a member of the Institution and Member of the Quantity Surveyors' Committee of the Chartered Surveyors' Institutions, acted as quantity surveyor for the six ventilation buildings for the tunnel. He was responsible for measuring up and quantifying the drawings for the individual ventilation stations. From this he calculated the material, build and labour costs for the ventilation stations, and his figures were issued to the various firms tendering for the work to enable them to submit their competitive estimates of cost.

Queensway ventilation diagram.
(Wirral Archives)

Wirral ventilation towers.
(Wirral Archives)

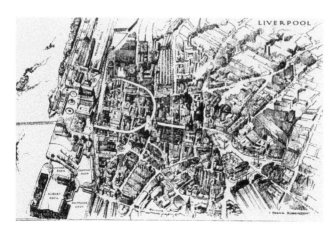

Liverpool ventilation towers.
(Wirral Archives)

Construction of tunnel ventilation shafts. (Birkenhead Library)

Bretherton buildings axis supply vent. (Birkenhead Library)

Brick assembly to supply shaft. (Birkenhead Library)

North John Street, Liverpool, exhaust chamber and tunnel roof prior to concreting. (Liverpool Museum)

Exhaust chamber and tunnel roof. (Liverpool Museum)

Fresh-air ducts land side. (Liverpool Museum)

Fresh-air ducts construction.
(Liverpool Museum)

Supply duct shuttering.
(Liverpool Museum)

Left: Walker fan blade. (Liverpool Museum)

Right: Sturdvent fan impeller. (Liverpool Museum)

St George's Dock building from the Strand. (Author)

St George's Dock fan. (Liverpool Museum)

St George's vent station emergency exit. (Liverpool Museum)

New Quay ventilation construction.
(Liverpool Museum)

Liverpool New Quay. (Author)

North John Street under construction.
(Liverpool Museum)

North John Street switch room. (Liverpool Museum)

North John Street today. (Author)

Sidney Street under construction. (Liverpool Museum)

Original Sidney
Street switch room.
(Liverpool Museum)

Above left: Sidney Street ventilation station. (Author)

Above right: Taylor Street ventilation station. (Author)

Above left: Woodside ventilation station. (Author)

Above right: Wirral ventilation buildings from Pier Head. This view is normally obscured by the Irish ferry at the terminal, shown by the five white posts. (Author)

Original Morpeth dock pump room. (Liverpool Museum)

Original mid-river pump room. (Liverpool Museum)

Architectural Features

Mr Herbert B. Rowse F.R.I.B.A was not only responsible for the six ventilation stations, he was also responsible for the entrances and approach portals along with toll booths, walls, lamps, glass lining and many more features both seen and unseen in the tunnel. The main entrances have arched pylons that are more than a decorative feature. They are utilitarian and the aesthetic is everywhere closely allied in the visible parts of the tunnel workings. Besides marking the limits of the approaches to oncoming traffic, these arches contain accommodation for the staff employed in the toll booths across the entrances.

The original toll booths were of cast-iron with a dado of decorative metalwork. The upper portion is in glass with opening sections, and lighting cove all round. These booths were emerald green and gold. Above the windows there is a symbolic design suggestive of speed. Dominating the area around the main entrances at Liverpool and Birkenhead are the lighting shafts, 60 feet (18.28 metres) high, constructed of reinforced concrete and overlaid with fluted and polished black granite. The shafts are surmounted by glazed bowls of gilded bronze with a decorative pinnacle above, serving to mark the entrance for approaching traffic. The idea was to provide adequate lighting for the area round the entrances, and at the same time to give some monumental expression to this great engineering achievement. The hollow columns contain ladders to provide easy access for maintenance purposes.

The Liverpool shaft was erected in reinforced concrete by Natal, and the Birkenhead shaft by McAlpine & Sons. The Trussed Concrete Steel Company supplied the steelwork, whilst the fabrication and erection of the granite was carried out by John Stubbs & Sons. The shafts are not merely tapered, but are slightly curved to correct the optical illusion of being a smaller diameter in the centre than at the top of the shaft. Every section of the granite was shaped individually to the corresponding position and curve of the shaft. Bronze tablets have been attached at the base of each lighting shaft with the centre tablet commemorating the Royal opening of the tunnel by His Majesty the King. Another records the names of the members and officers of the Mersey Tunnel Joint Committee. A third tablet incorporates the names of the engineers, the architect and the valuer, whilst the fourth is an acknowledgment of the services performed by the principal contractors.

The Portland Stone portals are surmounted by immense carvings designed by Thompson & Capstick, in association with the architect. Over the Liverpool entrance is an ornamented circular shield supported by two flying bulls of strikingly vital conception, and with a winged wheel above; all symbolic of swift and heavy traffic. The sculpture for the Birkenhead entrance is more formal, but no less striking. In this case a triangular shield is used, over which there appears the head of a figure suggestive of a motorist at speed.

Within the tunnel itself, a black glass lining ran the full length of the tunnel; now covered by the modern white panelling that is constantly damaged by high sided vehicles. The design and the finished appearance were equally simple and easily cleaned, as well as being easy to remove for repairs or replacement.

The lamp standards at the New Quay and Birkenhead entrances are made of special centrifugally spun concrete with a white Portland stone finish. These were manufactured by the Liverpool Artificial Stone Co. Ltd. In addition to the lamp standards, an immense quantity of paving slabs and kerbing were also supplied by this company, who installed a special plant to meet the requirements of the engineers. The paving slabs were manufactured with British granite and cement, under hydraulic pressure of 600 tonnes. They had to be reinforced to withstand a specified load and given a special stone finish because they cover the conduits carrying power and telephone cables through the tunnel.

The ornamental cast-iron work to the central lighting shafts, the toll booths, and the parapet and wall lamps was carried out by Messrs. H. H. Martyn, whose work throughout displays the finest qualities of metal craftsmanship. The interior steel linings and equipment of the toll booths have been supplied by Messrs. Roneo Ltd. The electrical equipment of the toll booths, the central lighting shafts and the parapet and wall lamps was undertaken by Messrs. Higgins & Griffiths. The electrical clocks and calendars in the toll booths were supplied by the Synchromatic Time Recording Co. Ltd.

Polished Granite Podium

This is a feature that has proved an interesting one to visitors to Liverpool. It shows the traverse of the tunnel itself under the river bed, branching out into its four entrances, and also the positions and characteristic profiles of the six ventilation buildings, three on each side of the river.

The Glass Dado

The Marplax finish was the finish of choice of Mellowes Ltd of Sheffield, and Pilkington Ltd of St Helens – both of whom played a considerable part in deciding this choice of product. It is simple in design and of elegant appearance. It can be easily cleaned and the maintenance costs are practically nil.

The dado is made up of quarter-inch-thick (6 mm) sheets of black glass, laid in lead caves and framed with stainless-steel bars. The natural vitreous surface of the glass provides a perfectly durable lining, which is unaffected by moisture or fumes. The dado runs the full length of the tunnel to a height of 6 feet and 3 inches (1.9 metres) from the concrete plinth level, and consists of three graduated rows of glass in large sheets supported by special rails of non-ferrous metal. These rails are designed with a movable web at the front and with two contact cushions at the back, so that whilst the glass is held firmly in position, there remains sufficient room for slight movement due to contraction, expansion or vibration.

The Plazas

Each of the four entrances were planned by the tunnel engineers with the co-operation of the architect, and in consultation, on the Liverpool entrances with the City Engineer, and at Birkenhead, the Borough Engineer. Their construction has involved a considerable amount of thought as to their use and ease of allowing traffic to flow freely in and out of the tunnels. The scale and character of the plazas are worthy of the great engineering feat which the tunnel represents.

The Concluding Stages (1933–1934)

Work on the tunnel progressed at a steady rate until June 1934 and this is where we pick up the additional works which were to become the six new ventilation stations that were to cost in excess of £570,000 (£35,539,671.00). Time was not on their side, nor was the solution, as the dates were set for the tunnel opening, but the work on the tunnel was progressing nicely. In other areas of the tunnel work, the surfacing of the roadways with studded iron plates, the ornamental glass dado – 5 feet high – and the finishing touches to the great arches with cement and fine plaster were being made. At Christmas 1933, and Easter 1934, the tunnel was thrown open to the public, in the cause of charity, and nearly 300,000 people travelled through the main tunnel on foot and expressed their admiration for such a wonder of modern engineering.

As the public were getting their view of the tunnel for the first time, the large traffic areas at the Old Haymarket (Liverpool) and Chester Street (Birkenhead), as well as the two dock-side entrances, were under construction. Although construction may seem a progressive word, the reality of this project was that a large number of properties and businesses were moved to other locations to make way for the tunnel.

Water Supply

The supply of water required for the cleansing of the tunnel and the service of the hydrants situated at each of the fire stations has necessitated the installation of a 4-inch (100 mm) main, which runs throughout the entire length, and is tapped at convenient points so that water can be drawn as required. At the boundary line between Liverpool and Birkenhead a stop-valve was added. This valve will remain closed; the Liverpool and Birkenhead sections of the tunnels were being supplied with water by the respective Corporations. In the remote event of a failure of water supply at either end, the stop-valve will be opened to enable the whole length of the main to be fully supplied from the alternative supply.

Given the historic significance of the Queensway Tunnel, certain elements have been listed by English Heritage (now Historic England). Below is a list of the tunnel elements and their description as detailed by Historic England:-

Ventilation Station of the Mersey Road Tunnel, Pacific Road

Ventilation station for the Mersey Road tunnel. 1925–34. Sir Basil Mott and J. A. Brodie, engineers, Herbert Rowse, architect. Steel framed with brick cladding. Composed of a series of geometric blocks grouped around giant main tower. Decoration provided by nogged projecting brick courses on main tower in the form of a cross, the tops of each block stressed with bands of stepped and nogged brick courses. Brick bands give a rusticated effect to base. Doors with chevron decoration in tower base. Houses giant fans used in ventilation, the largest of a series of three towers on the Birkenhead side of the tunnel

Ventilation Station of the Mersey Road Tunnel, Sidney Street

GV II Ventilation Station to the Mersey Road Tunnel. 1925–1934. By Sir Basil Mott and J. A. Brodie, as engineers and Herbert J. Rowse, architect. Brick faced, steel framed construction. Massive twin towers, largely blind but with double doors in south elevation divided by three blind lancet slits. Doors have chevron bands and architraves formed of brick in rippled courses, with central wing motifs. Ribbed brick quoins to towers, blind lancets and geometrical ribbed decoration towards the top. Built to house fans as part of the ventilation system for the road tunnel

Ventilation Station of the Mersey Road Tunnel, Taylor Street

Ventilation Station of the Mersey Road Tunnel. 1925–34. By Sir Basil Mott and J. A. Brodie, engineers, and Herbert J. Rowse, architect. Brick faced steel framed construction. Tower comprised of a series of grouped geometric blocks, largely blind but with small doorway in eastern elevation, and double doors in northern elevation. Doors themselves original features, with bands of chevron and scallop decoration. Architraves formed from rippled coursing of brick, with wing motifs in centre. Some ornamentation in the brickwork, with diamond frieze at base and giant cross motif in the largest of the blocks. Chevron bands of brickwork mark the tops of the towers, and the largest block is marked by a frieze of recessed panels or slits. Built to house fans as part of the ventilation system for the road tunnel. (*The Buildings of England*: Pevsner N and Hubbard E: Cheshire: Harmondsworth: 1971–)

Former Dock Entrance to Mersey Road Tunnel, Rendel Street

Dock entrance to Mersey Road Tunnel, now disused. 1931–34. By Sir Basil Mott and J. A. Brodie as engineers, Herbert J. Rowse, architect. White Portland stone. Retaining walls marking approach each side of tunnel portal, with low terminal blocks at each end, and tunnel entrance.

Moderne style with Egyptian detailing. Tunnel entrance with low relief shield flanked by winged beasts over. Scalloped cornice to retaining walls built on curve to each side. Terminal blocks stepped in plan and battered in section. Narrow bands of window, and stone grid-work with ribbed cornice band with arrow motif in low relief

Entrance to Mersey Tunnel, King's Square

Tunnel entrance. 1925–34. Engineered by Sir Basil Mott and J. A. Brodie, with Herbert J. Rowse as architect. Faced in white Portland stone. Egyptian style. Two flanking lodge towers, and walls leading downwards towards portals. Lodge towers have round-headed arched entrances flanked by fluted engaged shafts, and wing motifs in low relief over. Scalloped frieze and high blocking course above. Flanking walls link to retaining walls of tunnel entrance, with chevron frieze. Wing motif over portals. Original height of westernmost lodge tower reduced during alterations which have included additional building against the flanking wall and partially glazed infilling of formerly open ground floor of towers

Monument to the building of the Mersey Tunnel, Chester Street

Monument commemorating construction of Queensway (Mersey) Tunnel. 1934. By Herbert Rowse. White ashlar base with polished black granite shaft. Tall fluted shaft on high base with chevron and acanthus decoration in bands at base and top. It is capped by a fluted glass bowl out of which a ribbed cap carries a banded sphere. Base is inscribed with names of the engineer and architect of the tunnel, Sir Basil Mott J. A. Brodie, engineers, Herbert Rowse, architect, together with the names of the construction teams and committees. The monument was designed to occupy an axial position in the tunnel approach but is no longer in its original position

Entrance to Mersey Tunnel, Old Haymarket

Entrance to tunnel, retaining walls and Lodges. 1925–34. Sir Basil Mott and J. A. Brodie with Herbert J. Rowse as architect. Portland stone. Originally an axial and symmetrical design now obscured by subsequent alterations to the layout. Lodges in the form of triumphal arches to left and right of principal axis and retaining walls to the entrance of the tunnel which is a broad segmental arch. Retaining walls and tunnel entrance have some Art Deco ornamentation and sculpture now partly hidden. The lodges are tall and cubic with fluted buttresses and more Egyptian Art Deco ornamentation

Mersey Tunnel Entrance, New Quay

Tunnel entrance. 1925–34. Sir B. Mott and J. Brodie Engineers with H. J. Rowse as architect. Portland stone. Tunnel month has decorative plaque over. Long curving wall to right has reeded capping and ends with plaque with winged wheel motif. Iron lamps standards along wall. Wall to left is shorter, with no features of interest

New Quay Ventilation Station to the Mersey Road Tunnel, Fazakerley Street

Tunnel ventilation system. 1925–34, Sir Basil Mott and J. A. Brodie with Herbert J. Rowse as architect. Built as part of the Mersey Road Tunnel. Brick block with two framed entrances and bands of brick decoration at top and bottom, and three ventilation slits. Two towers to rear with decorative brick work and copper coving. Original doors

Ventilation Station to the Mersey Road Tunnel, North John Street

Ventilation station. 1925–34, Sir Basil Mott and J. A. Brodie with Herbert J. Rowse as architect; built as part of Mersey Road Tunnel. Portland Stone. Enormous tower-like structure with design emphasising mass and vertical lines, housing giant ventilating fans. Set on massive rectangular base having windows in vertical bands with Art Deco ornamentation set in horizontal bands above. Tower contains enormously tall slender blind window flanked by attached reeded columns

Georges Dock Ventilation and Central Control Station of the Mersey Road Tunnel, Georges Dock Way

Ventilation Station and offices. 1925–34. Sir Basil Mott and J. A. Brodie with Herbert J. Rowse as architect. Built as part of the Mersey Road Tunnel. Portland stone storeys. Rusticated ground floor. Windows to north and south in tall recesses flanked by relief sculptures. West facade has windows to end bays, and entrances, centre block has entrance flanked by niches containing bronze figures, term over. East facade has three empty niches, paved areas to north and south have retaining walls, rails and lamp standards, two square piers to north have rusticated bases and banded caps

Casting moulds for the iron road sections. (Liverpool Museum)

Note the age of the worker and the lack of breathing apparatus to protect from the fumes of the sulphur. (Birkenhead Library)

Casting the iron road sections. (Liverpool Museum)

Tunnel units factory. (Birkenhead Library)

Laying the iron road sections. (Liverpool Museum)

Opening ceremony commemorative column and plaques today. (Author)

Above left: Opening ceremony commemorative column plaques. (Author)

Above right: Opening ceremony commemorative column plaques. (Author)

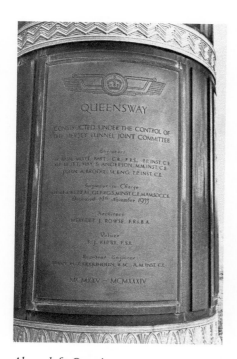

Above left: Opening ceremony commemorative column plaques. (Author)

Above right: Opening ceremony commemorative column plaques. (Author)

Composite tunnel entrance drawing.
(Birkenhead Library)

Birkenhead tunnel
approach *c.* 1934.
(Birkenhead Library)

Early days of tunnel in
use. (Birkenhead Library)

Birkenhead Tunnel entrance carving. (Author)

Liverpool (Haymarket) entrance. (Liverpool Museum)

Haymarket entrance, 2014. (Author)

Looking up William Brown Street (Liverpool Museum and Art Gallery) from above tunnel exit in Haymarket. (Author)

Queensway Tunnel Opening Ceremony

This thoroughfare is great and strange. The wonder of your Tunnel will only come into mind after reflection. Who can reflect without awe that the will and power of men, which in our time have created the noble bridges of the Thames, the Forth and Sydney Harbour, can drive also tunnels such as this, in which many streams of wheeled traffic may run in light and safety below the depths and turbulence of tidal water bearing ships of the world?

Many hundreds have toiled here and the work of many thousands all over the country has helped their toil. I thank all those whose effort has achieved this miracle.

May those who use it ever keep grateful thought of the many who struggled for long months against mud and darkness to bring it into being. May our people always work together thus for the blessing of this Kingdom by wise and noble uses of power that man has won from nature.

King George V, 18 July 1934

With these words the King, watched by between 150 and 200,000 people, activated the gold switch to raise the curtains covering the entrance to the Tunnel and officially opened one of the greatest feats of British engineering to the public. An undertaking which had been years in the planning and nine years in construction was finally complete.

An article appeared in the local *Liverpool Echo* the day before the opening, titled 'The Old Salt's Lament':-

"New Fangled Things – Like Tunnels." The reporter was talking to an old seaman who had worked on the river and at sea for all his life and was not too impressed with the "new fangled tunnels." He questioned the reporter by asking in his husky almost angry voice "I s'pose you don't see no beauty in that. I s'pose the likes of you sees only beauty in tunnels and them new fangled tunnels" as they sat on the railings of the ship he had just brought into Liverpool.

Waving his arm up and down the river he further stated "The whole lots picturesque, nothing like it anywhere and I've spent my whole life on it, I'ave" Further on in the discussion he noted "Them as goes motoring through this 'ere tunnel won't even know they have been in Liverpool. They won't! All they know is they went down a hole in one place and came out of a hole in another. This here river's Liverpool! If a town has got what they call a soul, this is it (pointing to the river) down here. St George's Hall isn't Liverpool, The town hall isn't Liverpool and the streets isn't either. You go around the world and every port you come to you'll see 'Liverpool' on some ship." Commenting on his using the tunnel the old seaman stated "not if the Lord Mayor and his own self goes on his knees to me, I won't go through it. Because, them what wants to burrow can burrow, but whilst I've my eyes and there's things to see on top, they won't get me to be no bloomin' rabbit."

Opening Ceremony

The day's programme was full and in order to do the monumental events justice I have listed them below, not only for the opening of the tunnel, but also the opening of the new main library in Birkenhead. This replaced the Carnegie Library, which was demolished to make way for the new tunnel.

10.50 a.m.
Their Majesties the King and Queen arrived at the city boundary – East Lancashire Road – and were met by the Lord Mayor and the Lady Mayoress (Councillor and Mrs George A. Strong). The Lord Mayor presented the Lady Mayoress, Sir Thomas White, the Town Clerk (Mr Walter Moon), and the Chief Constable (Mr A. K. Wilson).

Their Majesties the King and Queen left in the Royal Car, escorted by police on motorcycles. The procession passed along East Lancashire Road and onto Walton Hall Park in the following order:

Private Car
The Earl and Countess of Derby
Miss Ruth Primrose
Lord Stanley

Royal Car
THEIR MAJESTIES THE KING AND QUEEN
Dowager Countess of Airlie (Lady in Waiting) – The Countess was to serve Queen Victoria, Queen Alexandria and Queen Mary, before she died in 1956
Mr Leslie Hore-Belsha M.P. (Minister-in-Attendance)

Private Car
Lady Sefton
Lady Maureen Stanley
Lord Sefton
Major Hon. A. Hardinge (Assistant Private Secretary)

Private Car

Captain Bullock

Lieutenant Colonel R. H. Seymour (Equerry in Waiting)

11.10 a.m.

Their Majesties the King and Queen were received at Walton Hall Park by the Chairman of the Parks and Gardens Committee (Mr George Holme), who was presented by the Earl of Derby. The following were then presented: Councillor Peter Kavanagh, deputy chairman of the Parks and Gardens Committee, and Mr J. J. Guttridge, who was the chief superintendent and curator of the parks and gardens.

Walton Hall Park had been extended at a cost of £37,000 (£2,362,174.87) to include a spacious boating lake and boathouse for forty-eight boats, a two-acre model boat lake, two-storey open air café with verandas on the ground and spacious balconied above, all overlooking the lake. Also included was a bandstand, boulevards, walks, cricket, tennis, football and baseball grounds and a children's open air gymnasium. For adults there were the putting greens, bowling greens and tennis courts. The auditorium, which fronts the bandstand, can now be enlarged to cover a capacity of 15,000 people and used for open air plays, pageants, concerts and other presentations.

His Majesty the King duly declared Walton Hall Park open without leaving the car.

11.15 a.m.

The Royal procession proceeded via Walton Hall Avenue, Queens Drive, Muirhead Avenue, Derby Road, through Newsham Park by way of the Seaman's Orphanage to Prescot Road, Kensington, Prescott Street, London Road and William Brown Street and the length of Kingsway to the entrance of the Tunnel.

As the Royal Party made their way down London Road, the crowd of 150 to 200,000 people joined in to sing the first verse of Elgar's 'Land of Hope and Glory', accompanied by the band of the Liverpool City Police. The singing turned to murmuring as everyone stood on tiptoe to get a view of the King and Queen as they went past. Some of the crowd had prepared themselves for a long day and had brought provisions for at least one meal, and it was said that the scenes were reminiscent of Armistice Day. The most striking form of decoration was that formed by a group of over a thousand school children who were dressed in vivid coloured frocks and hats and had been carefully grouped, tier upon tier, on the steps of the museum in William Brown Street. The sight of this display formed a human floral bouquet. This was all part of the planned day's events.

11.45 a.m.

A portion of the initial route to the tunnel was lined by the Royal Naval Volunteer Reserve (Mersey Division) and the 55th (West Lancashire) Division (Liverpool) along with the British Legion, now the Royal British Legion.

Their Majesties the King and Queen were received by the Lord Mayor and Sir Thomas White, the Chairman of the Mersey Tunnel Joint Committee. Lord Derby, as Lord Lieutenant of Lancashire, presented The General Officer Commanding Western Command (Lt General Sir Walter M. St. G. Kirke) and the General officer Commanding West Lancashire Area and 55th (West Lancashire) Division (Major General W. J. N. Cooke-Collins).

His Majesty the King inspected the Guard of Honour mounted by the 2nd Battalion, The Kings Regiment (Liverpool) and simultaneously a detachment of the West Lancashire Territorial Army Nursing Service was inspected by her Majesty the Queen, accompanied by the Countess of Derby and the Lady Mayoress.

A procession was then formed as follows, whilst the Royal Party and the dignitaries made their way to the dais for the official opening ceremony:

Civic Regalia
The Town Clerk, Sir Thomas White
The Lord Mayor
THEIR MAJESTIES THE KING AND QUEEN
Mr Leslie Hore-Belisha, M.P., The Minister-in-Attendance
Lord Derby
The Ladies and Gentlemen in Waiting

Upon their Majesties taking their seats, the Lord Mayor tendered the city's welcome and asked their Majesties' gracious acceptance of an address from the Mersey Tunnel Joint Committee. Sir Thomas White (Chairman of the Mersey Tunnels Joint Committee) read the address and handed the address to his Majesty the King upon completion.

Sir Thomas White read:-

To the King's most excellent Majesty,

MAY IT PLEASE YOUR MAJESTY,

We your loyal subjects, the Members of the Mersey tunnel Joint Committee beg leave to approach your Majesty and her Majesty The Queen with an expression of our dutiful homage and our gratification at the high privilege of welcoming your Majesties on this the memorable occasion of your visit to Liverpool and Birkenhead, and we assure your majesties of our deepest loyalty and affection.

It is a matter of great gratification to the citizens and burgesses that your majesties have once more evinced deep interest in all that concerns the welfare of your people by your presence here today for the purpose of opening to public traffic the largest sub aqueous tunnel in the world – a great engineering work of natural character and importance which has been constructed to meet the ever growing transport requirements of the Port of Liverpool and an event of great importance to our trade and local history.

Your Majesties' presence on this eventful occasion is regarded by the citizens and burgesses of Liverpool and Birkenhead as another example of the profound interest taken by Your Majesties in all great projects having for their object the welfare and prosperity of Your Majesty's subjects.

It is our heartfelt prayer that the Almighty may continue to bestow every blessing upon Your Majesties and that Your Majesties may both be long spared to this great empire to advance the happiness to which your best endeavours are always directed.

Given under the common seal of the Mersey Tunnel joint Committee this eighteenth day of July One Thousand and Thirty Four.

His Majesty the King then read his reply and officially declared the tunnel open:-

Thank you for your address to the Queen and myself. It is with deep pleasure for us to come here today to open for the use of men a thoroughfare so great and strange as this Mersey Tunnel now made ready by your labour.

In some other seaport channels and estuaries have been bridged with structures which rank among the wonders of the world. Such bridges stand in the light to be marvelled at by all. The wonder of your tunnel will only come into the mind after reflection.

Who can reflect without awe that will and power of men, which in our own time can create the noble bridge of the Thames, the fourth, the Hudson and Sydney harbour, can drive also tunnels such as this, wherein many streams of wheeled traffic may run. In light and safety below the depth and turbulence of a tidal water bearing the ships of the world?

Such a task can only be achieved by the endeavours of a multitude. Hundreds have toiled here, the work of many thousands all over the country has helped their toil. I thank all those who efforts have achieved this miracle.

I praise the imagination that foresaw the minds that planned, the skill that fashioned, the will that drove, and the strong arms that endures in the bringing of this work to completion.

May your peoples always work together thus for the blessing of this kingdom by wise and noble uses of the power than man has won from nature.

I trust that the citizens of this double city, so long famous as daring traders and matchless seamen, may for many generations find profit and comfort in this link that binds them.

I am happy to declare the Mersey Tunnel open. May those who use it ever keep grateful thought of the many who struggled for long months against mud and darkness to bring it into being.

The King declared the Queensway Tunnel open by operating an electrical switch which simultaneously indicated at the Kings Square Entrance in Birkenhead that the tunnel was now officially open.

Whilst the King was giving his response, Sir Thomas White and the people of Liverpool and Birkenhead listened intently to the King's reply through the various speakers positioned around both areas. His Majesty pressed the switch, which then raised the two yellow poles – one on either side of the entrance to the tunnel – and the emerald-green curtains – which had been draped across the tunnel entrance – parted, and each half of the tunnel entrance was revealed. As the curtains raised they revealed a message positioned across the entrance, which in red letters spelled out: 'Merseyside welcomes Your Majesties'.

As the national anthem played and the curtains began to rise, few were aware that the electrical mechanism had failed and instead two men were stationed either side, raising the curtains with hand cranks. They did such a good job of opening the curtains in synchronisation that no one noticed.

To commemorate the event, 150,000 local children were awarded medals and the city held a week of celebrations which included a Ceremony of Remembrance at the cenotaph at St George's Hall (opposite the Liverpool Tunnel Entrance and behind St Johns Gardens) in the city centre.

At the time of its construction, the Queensway Tunnel was the longest underwater tunnel in the world; a title it held for twenty-four years until the Kanmon Strait Tunnel – 200 metres longer – was built in Japan.

The Chairman of the Mersey Tunnels Joint Committee requested if His Majesty the King would graciously accept a small model of the tunnel entrance, and a Georgian Silver Cup was given to Her Majesty. The Police Band then played the hymn 'All People That On Earth Do Dwell' and the Lord Bishop of Chester offered The Lord's Prayer, followed by the Lord Bishop of Liverpool offering the benediction.

12.15 a.m.
The King and Queen left the dais to enter the Royal Car and the first verse of the National Anthem was sung, led by the band of the 2nd Battalion the Kings Regiment (Liverpool). Their Majesties then drove through the tunnel to Kings Square, Birkenhead, and stopped at the dais in front of the stand.

12.25 p.m.
With the Royal procession expected at any moment, a boy scout, who acted as a guard of honour in the tunnel entrance, fainted, and he was quickly carried away by his colleagues to receive first-aid from one of the many first-aid posts positioned around the area.

When the Royal Car was being driven up the incline to the Birkenhead exit, the territorial unit lowered their colours – a unit flag depicting battle honours – whilst at the same time the Royal Standard was opened and displayed at the head of the slender flagpole immediately above the tunnel cutting.

This was a signal for the cheering crowd, which was begun by school children as they had the highest and best view of the dais events.

The band of the His Majesties Grenadier Guards were posted at Kings Square along with the Colour Party and an escort of the 4th and 5th battalion of the Cheshire Regiment, as well as detachments of the 2nd Cheshire Field Squadron, Royal Engineers and the British Legion.

The Mayor made the following address of welcome from the Corporation:-

To the King's most excellent Majesty,

MAY IT PLEASE YOUR MAJESTY,

We, the Mayor, aldermen, and Burgesses of the County Borough of Birkenhead, desire to offer a loyal and hearty welcome to your Majesty and to Her Majesty the Queen on the occasion of your visit to Birkenhead in connection with the opening of the new tunnel and roadway under the River Mersey.

As Your Majesty is aware, the Corporation of Birkenhead have combined with the City of Liverpool in the promotion and completion of this important undertaking. It is our fervent hope that the new tunnel will not only provide facilities for the national well-being, but also assist in the revival of trade in this town.

Such prosperity as Birkenhead enjoyed in its past days was due principally to shipbuilding and the industries associated with the Birkenhead docks. The recent depression in these industries has been to us a matter for regret and concerns. We are happy to hear witness to the fortitude with which the attendant difficulties have been faced locally by Your Majesty's subjects. We therefore express the hope that this important date in in the town history may prove the beginning of a period of greater prosperity for Birkenhead.

We desire also respectfully to record our great satisfaction that Your Majesty has been graciously pleased to signify your willingness to open the new central library, which will provide a valuable addition to the public amenities of the borough.

We are grateful for the benefits derived from Your Majesty's reign of nearly a quarter of a century and we pray that under almighty God's guidance, our Majesty's may be blessed with good health and may long be spared to guide the destinies of a loyal and contented people.

The King said in reply,

I thank you, Mr Mayor, for your loyal and dutiful address, and for the cordial reception which you have given to the Queen and myself.

It is with great pleasure that I recognise the public spirit of Birkenhead Corporation in joining with the city of Liverpool in the construction of the Mersey Tunnel which is today being brought into use and I trust that an ample reward will attend this great venture which is promoting the flow of traffic should assist the return of prosperity.

You have taken a long view of future traffic needs, and your action in engaging upon this project in difficult times bears witness to your faith in the future of the Merseyside which has afforded in the past so many outstanding illustrations of civic and commerce enterprise.

I am confident that so long as that spirit of initiative prevails among your citizens no opportunity of advancing the welfare of your community will be neglected. In all such efforts I wish you Gods speed.

I am fully conscious of the difficulties and discouragement which the people of Birkenhead have so bravely faced in the past few years. The shipbuilding industry on which the prosperity of your city so largely depends has been one of most severely hit by worldwide trade depression, and it is my earnest hope that the improvement in employment in recent months will continue.

As the cars were about to return to the Royal dais, the Town Clerk could be seen rushing across to the side of the square nearest to the market. He returned with an elderly gentleman; Mr S. F. Gillingham, a Birkenhead Centenarian who was in his 102nd year. Mr Gillingham was assisted towards the Royal dais by a friend and the Town Clerk, but their King and Queen did not wait for Mr Gillingham to reach the dais. They left the dais and went to meet him. The crown cheered at this act of thoughtfulness and were touched by this single honour accorded to Mr Gillingham by the King and Queen, who was very grateful for their greeting. As the King and Queen returned to the dais, Mr Gillingham, despite his age, was seen to return to the crowd with a perfect, straight back from his very proud moment.

12.30 p.m.
As the Royal party left the dais, it passed a line of ex-servicemen who stood to attention, followed by a group of children who stood up and cheered enthusiastically as the Royal Car passed. The Royal Car proceeded along the three mile journey along Market Place South, Conway Street, Argyle Street, Grange Road, Whetstone Lane and Borough Road.

12.40 p.m.
At the new £60,000 (£3,002,278.48) Central Library, which had been designed by Messers Grey Evans and Crossley, the band of the His Majesties Coldstream Guards were positioned along with a detachment of the Chester Yeomanry under Major Williams, who acted as guard of honour flanked by members of the Birkenhead School and cadets from the training ship *Conway*.

To assist in keeping the crown entertained whilst they awaited the Royal party to arrive, the band of the His Majesties Coldstream Guards played. As the Royal Car approached the crown began to cheer and carried on for several minutes during the opening ceremony.

In front of the Library there was a stand which was topped by a crown and contained a silver-gilt case in book-form, decorated with the borough coat of arms in enamel. It was inscribed 'To commemorate the opening of the Birkenhead Public Library by His Majesty King George V, July 18 1934'.

The King pressed a silver switch in the cover of the book and a Union Flag over the door fell away and the doors swung open. The switch book was handed to His Majesty as a memento by Alderman G. A. Solly, the chairman of the Libraries Committee. The King expressed his thanks to Alderman Solly and his regret that he could not stay longer, as he was already behind schedule. The King also remarked that the souvenir switch case was very interesting and he thanked him for the gift.

This magnificent library was built in a classical style with bold pillars at its entrance and a symmetrical frontage. The building housed 65,000 books, along with a reference library, a reading room and a children's library. The façade is 156 feet long (47.5 metres), built of Portland stone and set back 70 feet (21.3 metres) from the existing pavement, which gradually rises to meet the library's main steps. It takes the place of the Carnegie Library that stood on the corner of Chester Street and Market Place South and that now houses the tunnel entrance. It was originally proposed to build a commercial college and a central school on the adjoining site. The college was eventually built a mile or so up the road as Borough Road College. This has now been replaced by a housing development.

12.45 p.m.
The Royal Car left and proceeded along Borough Road, Bedford Drive, Bedford Avenue, and then from Bedford Road to Rock Ferry Station.

12.55 p.m.
At Rock Ferry Station, a detachment of the Royal Naval Volunteer Reserve and members of the training ship *Indefatigable* were posted. The Mayor, Town Clerk, and Chief Constable were at the station until the departure of the Royal train.

Queensway Tunnel opening ceremony. (Birkenhead Library)

Commemorations for tunnel opening in Carnforth Street. Martha Rose is seen putting the finishing touches to the painting in the road. Her son George is seen holding the paint can. It is worthy of note that the painting is of a section of the tunnel, including the depiction of four lanes of traffic, as well as the proposed – but never carried out – tram section below. (Birkenhead Library)

Tunnel mosaic in Haymarket entrance. (Author)

Plaque reads: THIS REPLICA OF THE ORIGINAL MAP WAS CONSTRUCTED JULY 1994 TO COMMEMORATE THE DIAMOND JUBILEE OF QUEENSWAY TUNNEL. (Author)

Mid-river section of finished tunnel. (Birkenhead Library)

Today's tunnel (Liverpool Dock exit). Note new cladding system on walls. (Author)

Original tunnel wall at Liverpool New Quay (Dock) exit showing new cladding rail fixed to wall and awaiting new panel system. At some points in the main tunnel you can see damaged panels where large vehicles have gone too close to the wall and ripped the panel seam away. (Author)

Salt barn located at the maintenance area of the Queensway Tunnel. The flyover to the right of the picture is part of the 1969 scheme. (Author)

George Dock shaft cover. (Author)

Original toll booth. (Author)

Statue of King George
at the Old Haymarket
entrance. (Liverpool)
(Author)

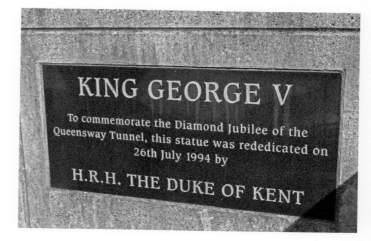

Plaque of Rededication
in July 1994 of the statue
of King George at the
Old Haymarket entrance.
(Author)

Statue of Queen Mary
at the Old Haymarket
entrance. (Liverpool)
(Author)

Plaque of Rededication
in July 1994 of the statue
of Queen Mary at the
Old Haymarket Entrance.
(Author)

Liverpool entrance commemorative plaque, seen to the right as you enter the tunnel. Plaque reads: IN COMMEMORATION OF ALDERMAN THE RIGHT HONOURABLE SIR ARCHIBALD T SALVIDGE P. C. K. B. E. FIRST CHAIRMAN OF THE MERSEY TUNNEL JOINT COMMITTEE 1925-1928 BY WHOSE VISION, FAITH AND COURAGE THIS TUNNEL WAS CONCEIVED AND CONSTRUCTED. (Author)

Liverpool entrance emblem. (Author)

King George's plaque for opening Birkenhead Library on 18 July 1934. Plaque reads: THIS LIBRARY WAS OPENED TO THE PUBLIC ON THE 18th JULY 1934 BY HIS MAJESTY KING GEORGE V, WHO WAS ACCOMPANIED BY HER MAJESTY QUEEN MARY. (Author)

Kingsway Tunnel

In early 1959 a local authority conference on cross-river transport discussed the options of either a new bridge over the River Mersey, or a new road tunnel under the river. It was decided that both options would be explored. The authority employed Megaw and Brown to undertake the report, along with W. S. Atkins and Partners, to undertake a limited traffic survey. The report looked at six tunnel schemes and two bridge schemes, with the favorite option falling for a six-lane bridge scheme. This was to be followed by a two-lane tunnel linking the existing branch tunnels. A tunnel could be built more economically, as the cost of a two-lane system compared to a double two-lane bridge would be of little difference.

The bridge scheme did not gain too many approvals as it proposed a 4,500-foot span (1371.6 metres), with six lanes. Following a number of meetings, a steering committee was set up in early 1962 to look at the options available. The committee was made up with representatives from the Cheshire, Lancashire, Liverpool, Birkenhead, Wallasey, and Bootle councils, as well as from the boroughs of Bebington, Crosby and Ellesmere Port. Representatives were also present from the Mersey Tunnel Joint Committee, the Mersey Docks and Harbor Board, the British Transport Commission, and from the Ministry of Transport. The committee may have taken heart that a new bridge on the other end of the river between Runcorn and Widnes had opened in 1961.

The name of the second tunnel was Liverpool's best kept secret, until it was finally revealed when Queen Elizabeth II pressed the button to release the long, blue curtains obscuring it. They fell away to reveal the 31-foot-long name (9.45 metres) – KINGSWAY, inscribed into Green Westmoreland slate, with gold leaf inlay. The opening was a simple yet colourful ceremony, watched by 6,000 people, which included 200 schoolchildren.

At the official opening on 24 June 1971, Her Majesty, Queen Elizabeth gave the following speech:-

I am delighted to be on Merseyside today, both in Liverpool and very shortly in Wallasey. The deep water of the River Mersey has been the foundation of Merseyside's growth. It has attracted shipping, encouraged trade, and helped to give Merseyside its own life and personality. It has been a highway to the world. But the river has also been a barrier. There has always been the problem of carrying people and goods across it. First there were ferries. These were later supplemented by the rail tunnel. Then in 1934 my grandparents, King George V and Queen Mary, opened the first road tunnel under the Mersey and the effect of the tunnel was dramatic.

But since 1934 traffic has grown enormously. Now one third of a million vehicles now pass through the tunnel every week.

The new tunnel is an essential part of the modernisation of the road network. It will give more room for crossing river traffic and it will make a valuable contribution to the life of both communities on both sides of the Mersey.

The construction of a tunnel of this size is always a notable feat of engineering. This one was not accomplished without conquering severe obstacles. The most modern methods have been used including the use of a mechanical mole in the excavation.

All this is a great achievement and Merseyside can justly be proud. I warmly congratulate the Mersey Tunnel Committee the architects, engineers, contractors and everyone concerned in this remarkable venture. My grandfather names the first Mersey Tunnel Queensway in honour of Queen Mary, so in honour of my father it is with greatest of pleasure that I declare the second Mersey Tunnel open and name it Kingsway.

Opening of second Mersey tunnel by Her Majesty Queen Elizabeth II, 24 June 1971

Order of Proceedings at Wallasey

2.35 p.m.
The Mayor of Wallasey – Alderman H. T. K. Morris – and the Mayoress, the Town Clerk of Wallasey – Mr A. G. Harrison, D.S.C. – and Mrs Harrison welcome the Lord Lieutenant of Cheshire – The Rt. Hon. Viscount Leverhulme, T. D., J. P. –, and Lady Leverhulme.

2.40 p.m.
Part of the proceedings at Liverpool were relayed over the public address system to Wallasey. These proceedings were as follows:

Her Majesty names the new tunnel and declares it open.

The opening is marked by a fanfare of trumpets at Liverpool and the firing of maroons from the Liverpool and Wallasey sides of the River Mersey.

The Lord Bishop of Liverpool dedicates the tunnel.

The Chairman of the Mersey Tunnel Joint Committee asks Her Majesty to accept a gift to mark the occasion – a silver model of the mole used to excavate the tunnel.

2.57 p.m.
The relay from Liverpool will side.

3.00 p.m.
Her Majesty leaves Liverpool and makes the first official journey through the new tunnel.

3.05 p.m.
Her Majesty arrives at the Wallasey Toll Area, accompanied by the Chairman of the Mersey Tunnel Joint Committee, Alderman H. Macdonald Steward.

The Queen is met by the Lord Lieutenant of Cheshire – The Right Honorable Viscount Leverhulme, T. D., J. P. –, who presented to Her Majesty the Mayor of Wallasey – Alderman H. T. K. Morris – and the Mayoress, as well as the Town Clerk – Mr A. G. Harrison, D. S. C. – and other distinguished persons.

The Mayor of Wallasey presents to Her Majesty the Member of Parliament for the Wallasey Constituency – The Right Honourable A. E. Marples, M.P. –, along with some leading members and chief officials of Wallasey Council.

The Mayor of Wallasey welcomes Her Majesty to Wallasey and invites her to unveil a plaque commemorating her journey through the tunnel.

Her Majesty unveils the plaque.

The Chairman of the Mersey Tunnel Joint Committee thank Her Majesty and invite her to inspect the administrative building and control room, as well as to meet representatives from the construction companies. The Queen also inspects the honour guard from the 1st Battalion Lancastrian Volunteers.

The tunnel is dedicated by the Bishop of Liverpool, the Right Reverend Stuart Blanch.

Although the second tube of the Wallasey tunnel was yet to be constructed, the tunnel was officially in use and, after years of planning and construction, the second river crossing was now complete. The second tunnel opened on 13 February 1974 with little fanfare and an estimated cost of £8.6 million (£92,943,321.80). The tunnel is a twin tube tunnel with two traffic lanes per tube and each tube is 3.65m in width. Prior to the opening there was a lot of intrigue as to the name of the new tunnel as it was shrouded in secrecy. Some of the suggestions included Dukes Way, Princes Way, Charles Way, and others which were a little more tongue in cheek; Mersey Mile, Link Way, Regal Way, Tunnel 2, Mersey Drive, Export Drive, Liver Drive, and Wallpool Tunnel.

In September 1963, the Mersey River Crossing Committee drafted an interim report recommending tunnelling rather than building a bridge. The tunnel was preferred as it offered greater flexibility in timing, traffic distribution and capital outlay, as well as a capacity for tidal traffic with 12-foot-wide (3.65 metres) lanes. If the need arose in the future, a tunnel could be duplicated with the capital outlay being spread over a longer period. This was to prove a major point, seeing as the river was a very busy port for all manner of goods and oil, and which would only get bigger – such as when Peel Ports expand the Seaforth Container Port in Liverpool, opening in late 2016.

Urgent preparation of a scheme for a new two-lane tunnel was put into planning. Consideration for using the old disused railway route from Seacombe was made for the possible Wallasey (Wirral) entrance to the tunnel. This railway used to take the public all the way to the Wallasey (Seacombe) ferry terminal, which then only had a two-minute walk from train to ferry. It also has Liscard and Poulton Stations within the cutting further up the line from the ferry terminal, close to the Mill Lane Bridge, with a connection with the present day Wirral line around the area of Bidston Station. Unfortunately passenger receipts were poor and the line closed in 1960, which resulted in it laying derelict until the Wallasey Tunnel Scheme arrived in 1966. Siting of the tunnel portal in Liverpool would be to the north of the proposed new inner ring road. The recommendations seemed favorable to the proposed scheme, which showed the most northerly of the six-lane tunnel proposals and a modified Liverpool with a re-planned inner-city road system. Parliamentary powers were requested for the construction of a tunnel and it was also recommended that Mott Hay and Anderson were to be the engineers. In November 1964, a bill was lodged in Parliament for the use of the Seacombe rail cutting, with the Liverpool portal coming about as a result of this.

Although this was strongly supported locally, the bill was not unopposed. As with the first tunnel, various councils fought against each other for their own respective views and for those of their constituents. Birkenhead preferred the tunnel linking with the existing branch tunnels, but Wallasey market gardeners opposed this as they objected to part of the proposed road layout.

The summer of 1964 was the start of preparations for the tunnel. This early start was to ensure the work would be carried out in a timely and efficient manner, following on from formal Parliamentary approval. This pre-contract and construction work included the preparation of the contract documents and all information and drawings necessary to form those contracts.

In 1965, during the passage of the Parliamentary Bill, a master program was prepared in anticipation of Royal Assent that summer followed by authority to proceed in the autumn, with a start on contract works at the beginning of 1966 with the receiving of Royal Ascent in 1968. Based on the preliminary designs prepared at the time, a construction period of almost five years was envisaged, with a completion of the project by autumn 1970; providing that the relevant authorization was obtained at the appropriate times and that there were no major unforeseen difficulties.

It was realized at an early stage that the acquisition of land, demolition of buildings, rehousing of displaced families and relocation of business would have to be phased beyond the start of work and, in fact, these operations took three years to complete. The early programming therefore provided valuable advance information on the priorities of acquisition and made allowance for approach road contracts to start with limited working sites, which increased in scope as more sites became available.

However, in June of 1969, Alderman Hugh Platt – leader of Birkenhead Council, and Deputy Chair of the Mersey Tunnel Joint Committee – asked if it was necessary to duplicate the tunnel. The latest figures showed that in April of 1969 a total of 1,599,953 vehicles used the existing Mersey Tunnel. This was 62,000 fewer vehicles using the tunnel than the previous year. The Mersey Tunnel Joint Committee was, however, expected to press for immediate duplication of the second tunnel following the publication of the Malts Report in late June of 1969.

Birkenhead Council was not expected to oppose the tunnel, but Alderman Platt questioned whether a second tunnel should be prioritised over the Merseyrail loop system. He went on to state that,

> In view of the latest tunnel figures, duplication of the second Mersey Tunnel is not the right priority. Once the second tunnel is opened it will only add around 20%, and this includes new approach roads, on both sides of the river, and a bypass in Rock Ferry.

Alderman Platt told the committee,

> The usage of the tunnels would continue to decline due to the saturation of parking in Liverpool. People are already seeking alternative means of crossing the river and the railway is the obvious one.

At this time the bill for the loop line had been passed in Parliament and the Chairman of the Mersey Tunnel Joint Committee, Alderman H. Macdonald Stewart, leader of Liverpool City Council, stated that,

> The twinning of the tunnels was always envisaged from the outset. I can't see anything in these monthly figures to suggest any difference and it could just be a poor Easter or something like that. I am taking no hard and fast line on this but we will have to see what recommendations Malts makes.

The sequence of construction and times of completion of the approach road contracts were considerably affected by the statutory services – the gas, water and electricity mains and cables, which had to be supported and diverted – and by the necessity of maintaining uninterrupted traffic flows across the works. The master program took account of these restraints and also gave a forecast of phasing of financial requirements during the construction period.

From the master program, more detailed programs for each contract were produced to form part of the contract documents, thus providing an overall control within which each appointed contractor was required to progress the work as it was allocated to the contractor. Each contractor was informed of the changing order of priorities as the work progressed, as well as the predicted completion of the project. In the event, the initial programmed progress on tunnel construction could not be sustained and was irretrievably delayed by labour disputes, as well as by both the TBM bearing breaking and geological problems found during construction. The actual construction period was five years and five months, with the tunnel opening on 24 June 1971 – less than six years from authorisation to the final completion of the tunnel.

The tunnels are circular in cross-section with an internal diameter of 9.63 metres and a painted, composite reinforced concrete and steel segmental provides the final finish for the road tunnel. Welded steel strips cover the lining joints between the segments.

A Robbins Tunnel Boring Machine – or 'TBM' – excavated an 11.2-metre-diameter tunnel; the majority of which was bored through the underlying sandstone rock formation. The South Tunnel was the first to be driven, and then the TBM was turned around and driven towards Liverpool. The TBM was known as the 'Mangla Dam Mole', after its previous work on the dam in Pakistan.

Under the River Mersey, the crown of the tunnel varies from between 7 metres and 15 metres. Cut-and-cover section approaches at either end of the tunnel were constructed through superficial soils and boulders. The South Tunnel has an oversized section of tunnel at mid-river that forms a 35-metre-long, 12.37-metre-wide and 10.76-metre-high emergency lay-by. This was constructed by enlarging the tunnel locally and lining it with an elliptical shaped cast-iron lining. The original configuration of the Kingsway tunnels incorporated links between the two road tunnels at the ventilation shafts, approximately 460 metres from the tunnel portals. Included in this was an additional link below deck-level at the mid-river emergency lay-by.

The path of the new tunnel was to be planned with the second tunnel next to it so as to enable its use during times of congestion. This would all depend on an adequate approach and the use of the mid-Wirral motorway (M53).

The new tunnels would enhance the importance of the industry on both sides or the river and the region as a whole. This was to prove a vital link in the war effort for the Second World War and the Atlantic Convoys. Vital supplies of men and materials came across the wolf-pack-infested Atlantic on a daily basis. This second Mersey Crossing would be designed to keep traffic away from the center of Liverpool and Birkenhead via the new ring-road and mid-Wirral M53.

The rock that would be cut through to form the tunnel was known to be reasonably uniform and sufficiently self-supporting so as not to require large amounts of additional support. However, the amount of water that was expected to be present within the tunnel during its construction through natural seepage from the rock was getting close to the maximum allowable for using a TBM. The use of a TBM allowed a smooth bore for the tunnel rather than a rough bore from explosives excavation. This second method would have additional works attached to it in that the rough edges would have to be smoothed out to allow a good seal with the tunnel walls. The TBM would alleviate this and allow the tunnel wall to be smoothly adhered to the tunnelled sections.

This development and possible method of construction was to be left open in the tender documents, and when the tenders were returned it was obvious that the companies that were asked to tender had chosen conflicting options. However, it was specified that tenders for the tunnel should be for excavation by mechanical means without the use of explosives, except as explicitly authorized. The lowest tender was for the normal method of construction and the next lowest had gone for a TBM method of construction. The choice of who was to be the successful tenderer came down to cost and time, and this would prove to be a difficult call.

The type of wall lining was now to be a consideration as cast-iron was expensive and difficult to lay, whereas concrete was cheaper and would prove to be an easier option. The decorative finish to the concrete would prove to be cheaper than the false lining required for the cast-iron sections. The sections, much like the cast-iron, could be manufactured in advance off-site if necessary. Concrete would be used for the main section of the tunnel which was within the sandstone, and cast-iron for the Liverpool end that was a mixture of sandstone and boulder clay.

The Wallasey Approach was to use the existing, but disused, Seacombe Railway cutting. This was clear of housing and businesses so there would be no need to compulsorily purchase property and relocate any affected businesses or residents. There were disadvantages to this, however, in that the width of the cutting was less than required by the four-lane approach road. The curves of the railway were also considerably less than those of required for motorway traffic, and the over bridges were lower and much narrower.

On the Liverpool Approach, the filled-in section of the Leeds and Liverpool canal, as well as the old abandoned Waterloo Dock Goods Yard, was available for clearance and demolition. This would allow the exit and approach to miss central Liverpool and proceed along its way to the proposed inner ring-road and out of the city. The exit to the tunnels on the Wirral side allowed for crawler lanes, allowing five lanes to aid the flow of fully-laden heavy-goods vehicles through the toll booths. It also allowed faster traffic to overtake, thus enabling the free flow of traffic through the tunnels.

A central lay-by was added to the tunnel at its lowest point to allow vehicles to be parked during busy periods and the tunnel to flow freely. The parked vehicle would be moved at a quieter period when tunnel traffic flow allowed one lane to be closed to facilitate the vehicle's removal. If the breakdown was enough to close the full tunnel, traffic would be diverted through the other tunnel. This would mean a little congestion as there would now only be one lane each way. An alternative solution would be to divert traffic through the Birkenhead Queensway until the obstruction had been cleared. This has happened on a number of occasions, and shows the flexibility of the construction by allowing traffic to traverse under the river Mersey to its final destination.

The toll area would be on the Wirral side as this would allow the existing practice of collection of tolls to proceed and thus avoid any unnecessary confusion in accounting. It would also allow the old railway goods yard of the disused railway to be used for the tolls and the entrance and exits from the tunnel. This area is 217 feet wide (66.14 metres) and provides a level space for the twelve toll booths.

The alignment, levels and gradients that were proposed for the new tunnel would be determined by the old Seacombe railway cutting, along with the riverbed and final point of exit from the tunnel. The tunnel was to be constructed by a Tunnel Boring Machine, at that time the largest in the world at 45 feet long (13.7 metres) and weighing 350 tonnes, with fifty-five cutter heads on the main cutting face to drive through the bunter sandstone. This would be a more efficient method than the hand-built tunnel of its predecessor. The tunnel lining was to be pre-cast, reinforced concrete with a welded steel inner face. Cast-iron segments were to be used where the tunnel rose out of the rock into boulder clay.

The road deck within the tunnel is a reinforced concrete construction with pre-stressed beams. Ventilation shafts and ducts connect the tunnel to a ventilation station on either side of the river. Drainage and sump pumps are positioned within reinforced concrete chambers. An area on the Liverpool side has a 50- to 60-foot (15.24 to 18.28 metres) cut and cover to form part of the tunnel. This cut and cover had a 100-foot span (30.48 metres), resting on a wall of 8-foot (2.4 metres) bored piles. The decorative finishes are of epoxy spray applied to the main primary lining where welded steel are used. Vitreous enameled steel-clad panels were placed above each walkway.

The Liverpool tunnel emerges completely from the rock along with the cutting. The Wallasey portal – the lower half of which is in sandstone – and the retaining wall virtually cover the self-supporting rock. Given the ground conditions the method of construction through the Triassic sandstone, boulder clay, peat, soft clays and sand was to be given a considerable amount of thought. In Bidston Moss (Wallasey approach and M53 connection), construction would have to contend with rock which had been eroded locally to 200 feet (60.96 metres), and which was covered in peat, soft clays and sand, and the River Fender Valley. The new tunnel would be approximately a mile further up river from the existing Queensway and be constructed in a more direct route across the river.

The new tunnels would have a diameter of 31 feet and 7 inches (9.62 metres) as an overall measurement. The two lanes in each tunnel would be 12 feet (3.65 metres) each, with a height of 16 feet and 6 inches (5 metres). The tunnel is a circular type with a small space under the roadway for a fresh air supply. The levels within the tunnel have a gradient of 1:30 (3.33%) on the Liverpool side and 1:25 (4%) on the Wallasey side, with vertical curves of 5,000 feet (1,524 metres) radius. The tunnel was to have a rock cover of better than 20 feet (6.09 metres). The opening portal positions of the tunnel are fixed at the lowest practicable road level. Liverpool has a depth of 58 feet (17.67 metres) and Wallasey a depth of 71 feet (21.64 metres).

The concrete precast segments were cast in the quarry at Penrhyn which is located near Bethesda in North Wales. At the end of the nineteenth century it was the world's largest slate quarry. The main pit is nearly 1 mile (1.6 km) long and 1,200 feet (370 metres) deep, and it was worked by nearly 3,000 quarrymen. The quarry was first developed in the 1770s by Richard Pennant and later by Baron Penrhyn, although it is likely that small-scale slate extraction on the site began considerably earlier. Much of this early working was for domestic use only as no large scale transport infrastructure was developed until Pennant's involvement. From then on, slates from the quarry were transported to the sea at Port Penrhyn on the narrow gauge Penrhyn Quarry Railway built in 1798; one of the earliest railway lines. In the nineteenth century the Penrhyn Quarry, along with the Dinorwic Quarry, dominated the Welsh slate industry.

The quarry holds a significant place in the history of the British Labour Movement as the site of two prolonged strikes by workers demanding better pay and safer conditions. The first strike lasted eleven months in 1896. The second began on 22 November 1900 and lasted for three years. Known as 'The Great Strike of Penrhyn', this was the longest dispute in British industrial history. In the longer term, the dispute cast the shadow of unreliability on the North Welsh slate industry, causing orders to drop sharply and thousands of workers to be laid off.

The tunnel segments were transported to Nutall-Atkinson's compound, on the site of the old swings. They were lowered to the portal area and transported to the TBM by single-cab Foden dump tricks. According to those in the tunnel at the time, the tunnel invert was not the safest place to be. The concrete test cubes were taken to the Sangberg and British Inspecting Engineers laboratory in Gorsey Lane, Wallasey, for testing. The laboratory also undertook the testing of the concrete on the Liverpool side and the Wallasey Approach roads.

The project for the second River Mersey Crossing could be seen as a more complex project than the first tunnel. Although the second tunnel is shorter, the links to it on both sides of the river proved to be large engineering feats in their own right. Work was to start in the autumn. Based on the preliminary designs, a construction period of less than five years was predicted, which was all down to the project receiving all relevant authorizations, and with the hope that no major unforeseen difficulties in planning or construction of the project occurred. Avoidance of noise during the tunnel's construction, particularly in built-up areas on the Wirral side – seeing as they were right next to the approach roads and tunnel route – was paramount. This would be achieved by using a TBM, which would also allow for a more accurate bore and placement of modular tunnel lining pieces, thus saving time and money on the construction.

A large scale project such as the Mersey Tunnel requires a large number of organizations, and specialist contractors. All of this has to be pulled together with one overall person or group in charge. With this project you have both tunnels and approach roads on both sides of the river; links with the Department of Transport for link roads and the central Wirral Motorway (M53) are required.

(a) The tunnel and approach roads were mainly constructed for the MTJC – the Mersey Tunnel Joint Committee – but the project also included some principal roads which were constructed for both the Wallasey and Liverpool Corporations, for whom the MTJC acted as Agent

(b) The viaducts and roadworks at Bidston Moss were constructed for the DoE – the Department of the Environment – as part of the trunk road plan. This contract was administrated by the Divisional Road Engineer for the North West on behalf of the DoE, who acted as Agent for the Wallasey Corporation in respect to the principal roads and for the MTJC in respect to the tunnel approach element

(c) The Consulting Engineers were acting on behalf of four clients, whose joint and separate interests were to be taken into account in the supervisory organization and procedures

The TBM used on the Wallasey Tunnels was a Robbins 371 which had been previously used on the Mangla Dam. This was the first of the two dams constructed to strengthen the irrigation system of Pakistan as part of the Indus Basin Project, with the other being the Tarbela Dam on the river Indus. US company Guy F. Atkinson & Co. of San Francisco sponsored the project for a consortium of eight American construction companies.

The TBM has a diameter of 36 feet and 8 inches (11.17 metres) and was the largest in the world, weighing in at 350 tonnes. To enable the TBM to be used in the Wallasey Tunnel, it had to be extensively modified and improved during the course of the construction of the tunnels. By the time construction was due to start, it was realised that the preferred TBC – which was used on the Mangla Dam in Pakistan – had been sold to a plant disposal company in Houston, Texas. The TBM had to be purchased and arrangements made for it to be shipped to Gravesend, arriving on 7 July 1966.

Once the TBM had passed through Customs and was transported to the work site, the modifications started. To enable this information to be ready for the TBM's arrival and subsequent use in the tunnels, a lot of work in San Francisco and London had to be completed.

The redesigned machine was erected and ready for moving forward to the Wallasey Portal face on 18 November 1967. After the last project, in dry conditions in Pakistan, there were detractors who said the TBM would not work in the wet conditions under the River Mersey, but they were proved to be wrong. Like all pioneers, the engineers and contractors pressed on with their endeavors and their courage and determination prevailed.

The TBM diameter was reduced to 33 feet and 11 inches (10.33 metres), with fifty-five cutters on the cutterhead. This was reduced in diameter to 5 feet and 6 inches (1.67 metres) to give access to the pilot tunnel. Alterations were made in the main assembly, as well as to the propulsion system and the provision of equipment for segment handling and erection. The rear cutting head was equipped to handle the concrete lining segments with a crane and placing boom. By adjusting the machine by a factor of 4 feet (1.2 metres), which was the width of one segment, a maximum of 16 feet (4.8 metres) of tunnel would be unsupported. This would be reduced to 12 feet (3.65 metres) when the concrete segment was in place.

The TBM would move itself through the digging, using the face it developed through the propulsion arms. These transmit the reaction to a system of thrust blocks, friction gripped on the tunnel lining at axis level, which spreads the load over twenty completed and grouted rings. The TBM was erected between two massive concrete walls, founded on bedrock outside the first portal, and anchored by a heavy horizontal beam extending forward at axis level to the tunnel headwall. The beams were fitted with thrust blocks along their length. This allowed the propulsion arms to the anchor walls until enough lining rings were placed to hold the tunnel lining.

Following extensive modifications for the wet and muddy conditions the TBM was set to work on a test run on 26 November with it starting to excavate the tunnel on 1 December 1968. By 10 February 1968, ring length number 47 was erected, and the tunnel had averaged around 30 feet (9.14 metres) per week. The distance covered was seen as a little disappointing and was partly put down to intermittent trouble with the tunnel crown. It was hoped that once they had passed ring 47 the problems would have subsided. However, by ring 53 it was found that the water table within the tunnel had risen approximately 6 feet (1.82 metres) above the tunnel's invert. This weakened the strata and a major outbreak of rocks began to rain into the tunnel. These conditions persisted with intermittent outbreaks until 7 March, when better rock was met at ring 83.

Another consequence of the problems was the damage to the TBM cutters. Large lumps of rock could be broken up by the miners in the pilot tunnel. However, with the TBM, the problems saw damage and even blockages to the muck chute and conveyors. This caused frequent stoppages and loss of production in the length of tunnel constructed by the day. Steel arch ribs and timber lagging were used to provide ground support in the pilot tunnel and these had to be removed at least 8 feet (2.4 metres) ahead of the TBM. This then allowed shoveling to take place. At the locations of the weak strata and rock falls, the steel ribs and timber arches were trapped and difficult to remove. The additional work to remove the ribs and lagging was not only hazardous but time consuming.

The sheer weight of the broken rock which was now sitting on the cutterhead caused concern for the tunnel management. One main concern was the possible damage to the main bearing seals and the cutterhead. On several occasions the cutterhead became trapped by the rock falls and due to the confined space access to the cutters for maintenance and inspection was not very practical.

The Liverpool Portal cofferdam was completed to its full depth in December 1967 and this allowed the placing of concrete for portal structures to start in January 1968. Work on the tunnel started in February and continued until the end of 1968. This continued on the premise that a high priority would be given to the stage for the main tunnel drive. By the end of April 1968 the special tunneling gantry, which was really a skeleton shield – known affectionately as the flying bedstead – had been delivered with its ancillary equipment, and erection of the forepart in the portal shaft had begun. When the completed forepart had entered the eye of the tunnel, the after part was erected in the shaft. The whole gantry could then move off as one unit, although the face had to be advanced over 45 feet (13.7 metres) before the gantry-tail left clearance for normal muck disposal and normal operation of the system. This stage was reached in mid-June.

The gantry platform travelled along the new tunnel on rollers which were fixed to the cast-iron lining. It was anchored to the lining via hydraulically operated sockets and propelled forward by rams. At the forward end it carried gun struts to support face timbering or decking, for hand excavation or drilling of rock for light blasting. It also had a roller type erector to place the lining over the upper half of the tunnel.

Excavation over the face was carried out by hand in boulder clay or sand, and by light charges of explosive in the rock. A Hymac hydraulic excavator was used to trim the lower part of the face, as well as to excavate and load away the waste material to Edbro lorries. All waste was hoisted up the portal shaft in the Edbro skips and tipped at the shaft head for distribution into the spoil heaps above ground. Grouting equipment was carried on the gantry, and the rings were grouted and caulked shortly after erection.

Unlike the Queensway Tunnel, the Kingsway Tunnel lining was designed so that it provides an almost finished tunnel. The concrete surface of the purpose-built lining-rings gives a substantial saving in cost and, along with the welded steel membrane backing, allows for a high standard of decorative finish to be applied. The main road deck construction started on 10 May 1970, and was completed on 7 February 1971.

The road deck construction was based upon the use of standard inverted T-section precast, pre-stressed concrete beams that were 24 feet (7.31 metres) long, 1 foot and 3 inches (381mm) deep and 1 foot and 8 inches (508 mm) wide. Mild steel reinforcement 40 mm in diameter was threaded through the steel beams and concrete was poured for the full width of the steel beams. Reinforcing mesh was then laid over the tops of the beams and 4 inches (100mm) of concrete was then placed over the beams. The concrete was sulphate resisting cement and the aggregates were granite from Penmaenmawr quarries, so Wales played a large part in the construction of the second river crossing.

By 1960, peak hour demand was exceeding capacity in the Queensway Tunnel, and as the entrances lead directly onto city streets, severe congestion was a serious consideration and growing problem. The main traffic flow from the countryside of Wirral to Liverpool was in the morning peak and vice versa in the evening. With only four 9-foot-wide (2.74 metres) lanes that could see 3,750 vehicles per hour, the tunnel authorities tried using three lanes in one direction at peak hours to assist the tunnel's flow.

The Liverpool City Engineer introduced a signal control system on the tunnel approaches with special queuing sections for waiting vehicles. This was further improved by the construction of a flyover for local traffic crossing Byrom Street, which was the main access to the tunnel. During this period, the Borough Engineer of Birkenhead was also responsible for the construction of tunnel approach flyovers linked with a large storage area, and in 1969 to 1970, all toll facilities were moved to the Birkenhead side.

To alleviate the existing traffic congestion in the center of both Liverpool and Birkenhead, the proposals for the new tunnel would be positioned to avoid this problem. This would be achieved in a former railway cutting for the Wallasey approach and a dedicated road access network on the Liverpool side. Both would pass under existing roads and with major junctions to the highway network at the extremities of the roads that were provided exclusively for tunnel traffic.

Wallasey has a two-mile section directly linked to the mid-Wirral motorway (M53) and the Bidston Moss approach road. Connections are also formed to the principal road system of North Wallasey and Birkenhead as well as access from the Wallasey and Birkenhead Docks area. An additional grade separated intersection at Gorsey Lane allows Wallasey traffic to travel to and from the tunnel, as well as the M53. This access point is located just outside the toll booth plaza, but does not provide for local traffic access. The approach at the Bidston Moss intersection allows traffic to traverse the tunnel. In the event of the Kingsway Tunnel being closed due to a major accident or for maintenance purposes, this is consequently the last diversion point to the Queensway Tunnel. This 'point of no return' defines the limit of the Mersey Tunnels' financial maintenance and policing commitments.

On the Liverpool side there is a large trumpet junction to Scotland Road, which was widened and improved in the late 1970s to urban motorway standard and which will provide free flow access southwards to the Liverpool Inner Ring Road. The approach roads are designed to allow a good traffic flow between Leeds Street and Scotland Road in Liverpool and the toll area in Wallasey. The access roads have a traffic capacity in excess of the tunnels and at peak times the Wirral-bound tunnel has two-way traffic to reduce any congestion at the tolls.

The final stages of both approaches are sunk below ground-level with retaining walls on both sides. The depth and details of the retaining walls was dictated by the tunnel portal levels and the topography of the sites. The retaining walls in the entrance to the tunnels reduced noise and were better than elevated roadways, which would be a blight on the overall skyline and surrounding buildings.

The Liverpool Approach road has a gradient of 3.33% over a distance of approximately 3,880 feet (1182.6 metres), with 1,300 feet (396 metres) of the cut-and-cover tunnel cutting. Vehicles leaving the tunnel via the west slip road must travel approximately 2,800 feet (853.4 metres) before they can use the full width between the twin tunnels that provides an additional climbing lane for heavy traffic between the portal and the west slip road.

The tunnel approach and exit layout is governed by the limitations of space, controlled on the eastern side by the realignment of Great Homer Street. This is a principal arterial route to the city with a dual 24-foot-wide (7.3 metres) carriageway designed for a speed of 30 mph, at-grade intersections. The tunnel approach forms a trumpet intersection with Scotland Road, which is widened to provide a 36-foot-wide (10.97 metres) dual carriageway. The slip roads are 24 feet wide (7.3 metres) with 8-foot (2.4 metres) hard shoulders and 3-foot (914mm) paved strips, allowing for a minimum separation on the offside of 5 feet (1.52 metres). The space between retaining walls in the dual carriageway trumpet section has been laid out with 11-foot (3.35 metres) and 5-foot (1.52 metres) hard shoulders for in-going traffic respectively. The 11-foot (3.35 metres) hard shoulder has been designed as a third lane for tidal flow working after tunnel duplication.

Wallasey Tunnel Dates

1963
2nd November – Committee decide on a new tunnel.

1964
15th February – Seven local authorities discuss idea.

27th November – Tunnel Bill before Parliament.

23rd December – Wallasey and Liverpool electors approve tunnel.

1965
6th August – Tunnel receives Royal assent.

1966
13th January – Lord Mayor of Liverpool Alderman David Cowley starts drill at 11.35 a.m. Mayor of Wallasey – Alderman C. S. Tomkins – starts drill at 12.05 p.m.

10th March – Vertical shaft at Liverpool 80 feet (24.38 metres) and Wallasey at 60 feet (18.28 metres).

1st April – Demolition of houses for approach roads.

31st May – Pump breakdown.

14th October – Pilot tunnel partially flooded after fire at pumps. Workmen fought to hold 750,000 gallons of rising water.

18th October – Normal work resumes.

15th December – Pilot shafts 700 feet (213 metres) apart.

1967
17th January – Breakthrough.
31st October – Engineers arm mole cutters at Wallasey.

2nd December – Mole starts work.

1968
24th June – Day and night shifts drive mole 212 feet and 6 inches (64.77 metres), which was claimed as a new world record.

19th November – Major flaw in river bed halts mole.

31st December – Mole starts again.

1969
27th February – Water and silt cause mole thrust to stop working.

17th March – Underground repairs to mole set up an engineering feat with the bearing.

22nd May – Mole starts upward climb.

16th October – Civic Leaders walk – or rather paddle – through from Liverpool to Seacombe and meet the mole on the way through.

1970
4th March – Breakthrough at Liverpool.

Once the mole had finished its work it was dismantled and put into storage on Waterloo Dock near Great Howarth Street, close to the tunnel dock exit. Assessors looked at the viability of the machinery and possible restoration. The costs of restoration would have been approximately £100,000 (£1,436,870.00), or scrapping it gave a value of £5,000 (£71,843.50). The only other option was to purchase the mole for £15,000 (£215,530.50) and make a monument out of it.

In an article in the *Liverpool Echo* in May of 1972, there was a discussion as to the viability of using the mole for a monument (Mott Hay Anderson) or even giving the mole to an underdeveloped country. The mole was the holder of several world records during the construction of the Kingsway tunnel; one record of which was a drive on tunnel two for a length of 7,400 feet, (2,255.22 metres) in fourteen months.

Wallasey Tunnel Statistics

Portal to Portal – Approximately 1.5 miles
Length including approach roads – Approximately 5.5 miles
Width of river – 1,000 yards
Diameter of pilot hole – 12 inches (horse shoe) (305mm)
Diameter of main tunnel – 30 inches (circular) (762mm)
Depth below river – 20 feet (6.1 metres) minimum or 40 feet (12.2 metres) average
Gradient – 1:25 max
Road width – 2 x 12 feet (3.65 metres)
Headroom – 16 feet and 6 inches (5 metres)
Lining – Mostly concrete segments / cast-iron
Ventilation – One each side

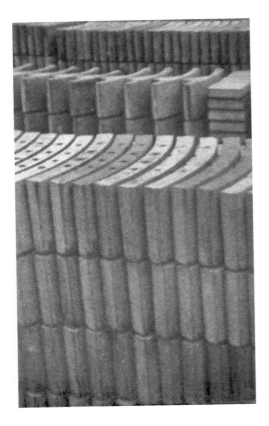

Tunnel segments awaiting delivery to site.
(Birkenhead library)

The following are general construction pictures of Kingsway Tunnel, some of which have specific titles (Birkenhead Library):-

THIRD MERSEY TUNNEL PILOT B
MINI — MOLE
DESIGNED & MADE BY
ROBERT L. PRIESTLEY
OF GRAVESEND KENT. FOR
NUTTALL ATKINSON

Mini mole machine.

Main mole machine cutter head.

Main mole machine.

Break Through.

Left: Mole recovery shaft.

Right: Liverpool Drive – Cement pallets on gantry.

General tunnel construction.

Left: BPL 2541 Kingsway Tunnel general construction.

Right: BPL 2542 Kingsway Tunnel general construction.

BPL 2543 Kingsway Tunnel general construction.

BPL 2546 Kingsway Tunnel
general construction.

BPL 2547 Kingsway Tunnel
general construction.

BPL 2548 Kingsway Tunnel
general construction for
access ramp to outside
construction area.

BPL 2549 Kingsway Tunnel
general construction.

BPL 2552 Kingsway Tunnel
general construction –
demolition of property for
new tunnel.

BPL 2553 Kingsway Tunnel
general construction.

BPL 2554 Kingsway Tunnel general
construction – cutting head.

BPL 2555 Kingsway Tunnel general construction.

General tunnel section.

Tunnel slurry containers being taken to surface.

Hymac excavator at work in Wallasey Tunnel.

Slurry in tunnel invert, October 1968.

Above left: Nuttal site wagon.

Above right: Liverpool sun visor under construction.

Offloading cement bags from Edbro lorry.

Slurry invert, October 1968.

Slurry removal.

Tunnel on completion of
cover strips to segments.

Tunnel
almost
complete.

Left: Tunnel worker with waterproof black coat.

Right: Roy – one of the many tunnel workers on site.

Despite being underground, an accurate survey for the direction of the tunnel dig had to be maintained.

Working conditions.

Placing walkway segments.

Completed tunnel.

Managers survey the completed tunnel.

Second tunnel drive team.

Kingsway Tunnel Ventilation

Unlike the Queensway Tunnel, the second Mersey Crossing was to have one ventilation building on each side of the river. One was to be a short distance from the Seacombe Ferry Terminal in Wallasey, and the other within Liverpool's old railway goods station in Waterloo. The site of the Liverpool building has a large wholesale building a short distance along the road, and the Old Waterloo Building is now apartments. Both sites and buildings are easily distinguished due to their individual design, unlike the Queensway ventilation buildings.

The ventilation buildings are positioned on either side of the river in a direct line with each other and as close to the river as possible. The Seacombe building has a guide along the promenade showing the line of the tunnel. The system was designed by Dr B. R. Pursall as a consultant to the committee. Each station draws fresh air into the tunnel via a horizontal shaft into the invert of the tunnel. Foul air is exhausted from the crown of the tunnel and discharged through the top of the ventilation stations. The towers discharge at a height of 162 feet (49.3 metres) on the Liverpool side and 185 feet (56.38 metres) on the Wirral (Seacombe) side.

Vent shaft. (Birkenhead Library)

Liverpool vent station. (Author)

Liverpool vent station. (Author)

Wallasey vent station.
(Author)

Line of tunnel marked along
Wallasey promenade, in
front of ventilation tower.
(Author)

Second World War

Between 1939 and 1945, the port of Liverpool handled more than 75 million tonnes of cargo – mostly from the Atlantic Convoys. 56 million tonnes were imports vital to the war effort and feeding the public, along with 18.5 million being sent back out as war supplies to the battlefronts throughout the world, including North Africa and mainland Europe. At the start of the war, many proposals were being considered for storing and using materials for the war effort, and one of those was the Mersey Tunnel.

To allow for all this work, the War Office took great interest in the Mersey Tunnels. In a letter dated 24 August 1940 from Sir William G. Chamberlain of the Ministry of Transport Office in Manchester to Sir John Reith GCVO, GBE MP of the Ministry of Transport in London, the Manchester Office reported on the survey of the Mersey tunnels to ascertain their possible use for the war effort.

It was in the early stages of the Second World War that the Queensway Tunnel caught the attention of the Ministry of Supply, as well as the Ministry of Transport. Secret correspondence was sent from the wartime ministry and the Liverpool City Engineer to see if the tunnel could be utilised for the war effort. William Chamberlain from the Ministry of Transport discussed the matter with the City Engineer, who was quick to point out that Liverpool was the busiest port in the Empire, predominantly due to it being the main gateway between America and Britain. It also had the statistic that most of its substantial cross-river traffic went through the tunnel. In addition to this, many of the old steamer ferries had been decommissioned shortly after the new road opened.

German view of Liverpool. Note bomb damage close to the docks. (Wirral Archives)

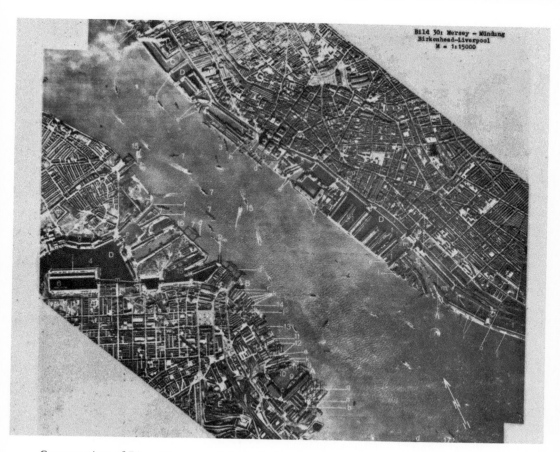

German view of River Mersey and docks in Liverpool and Wirral. (Wirral Archives)

1969 Changes to Tunnel Entrance

Prior to the mid-1960s there were around 3.3 million vehicles using the Queensway Tunnel and the traffic flow into and around the tunnel was now starting to back up in the surrounding areas and town centres. This was also exacerbated by the commercial and retail areas of Birkenhead becoming grid-locked due to the queue of traffic waiting to access the tunnel. In 1964, Mr H. C. Oxburgh drew up plans for a large scale concept of traffic management to assist in alleviating the congestion suffered in Birkenhead, especially during peak hours. The solution was to start the concrete mixers and start drawing up new flyovers to go at both ends of the tunnel. It was impossible to miss the fact that a whole new tunnel was being built to the north to relieve the Queensway, but that doesn't seem to have stopped a remarkable scheme to demolish huge areas of Birkenhead. By the early 1960s, it was becoming clear that the Queensway Tunnel was carrying much more traffic than had ever been envisaged when it was first opened. In just thirty years it was thick with cars and lorries throughout the day. The problem came in two parts: the capacity of the tunnel – which was set at four traffic lanes – and the bottlenecks along its route, and the toll booths at each entrance.

The proposed plan would allow different sections of traffic to flow a pre-determined route to the tunnel, local town traffic and local dock traffic to allow them to reach their destination. A system of viaducts and underpasses would separate the traffic and two automatic marshalling areas would accommodate the tunnel traffic. A bill was proposed in Parliament at the beginning of 1965 and it received Royal Ascent on 5 August 1965. It was known as the Birkenhead Corporation (Mersey Tunnel Approaches) Act 1965.

In addition to the powers granted in the 1965 Act, the Corporation also implemented the compulsory purchase of land and property under the Chester Street Clearance Area Compulsory Purchase Order 1961. This allowed 179 houses, flats and private homes, as well as 90 shops, 23 factories, workshops and yards, 14 public houses and a variety of other public buildings and land to be acquired. A large number of those affected by the order, especially local factories, were found new premises – many of which were located on land owned by the corporation – and 224 families were found new homes at a cost of £1 million (£15,144,600.00) for the whole operation.

Once in the marshalling area, the traffic would be controlled by feeding the lanes into the manned toll booths so that vehicles would be fed into the tunnel in an orderly manner, especially during peak times. In the early days of the marshalling yard, drivers were requested to turn off their engines and await their turn to proceed to the toll booth and then on to the tunnel. The system was designed to allow 4,500 vehicles per hour through to Liverpool, with an option to adjust the system's needs as required, such as if there were a lesser or greater number of vehicles wishing to travel to Liverpool. Vehicles were directed to the correct queue by an overhead traffic light system positioned on a gantry above the access road. Smaller signs were located at other access points and some even had what were described as 'Secret Faces', which would be illuminated when a diversionary route was used to ease the build-up of traffic in an emergency.

On approaching, the motorist could take any lane showing a white arrow at the entrance and join the queue in that lane. The exit from all lanes was blocked by red crosses. The computer system would then allow a small number of lanes to proceed by showing them a white arrow, and at the same time block the entrance to those lanes with a red cross. Once a lane was empty it was allowed to fill up again from the back and other lanes were allowed to proceed to the tolls.

There was an simple flaw with this scheme, which was that any regular user would quickly realise that the lanes were released in a sequence and would learn to join whichever lane was next to be released. The less scrupulous could always simply join an open lane and move across into one being released.

The system was installed in 1969–70 but never really worked properly. The computer system failed at its job quite miserably, and it was rapidly becoming clear that the new Kingsway Tunnel to the north was clearing many of the Queensway's problems. By the time the queuing system was built and active, the epic congestion it was designed for had gone.

The Tunnel Approach Scheme was opened on 15 July 1969 by Alderman Hugh Platt O.B.E., J.P. Chairman of the General Purpose Committee. His speech was as follows in a special brochure commissioned for the opening commemoration:-

It gives me great pleasure to mark the opening of the great engineering project by recording this message in the brochure specially commissioned to commemorate the occasion. We are now nearing the culmination of one of our efforts to resolve Birkenhead's serious traffic problems, and it highlights the results which can be achieved by hard work and co-operation between public bodies such as the Corporation the Mersey Tunnel Joint Committee and their staffs, the Consulting Engineers and the Contractors. They have all worked together with a single objective – to get the job done on time. It is only just over two years ago that the then Minister of Transport performed the ceremony to start the work. There is also an historic connection in the 35 years ago almost to the day (18th July 1934), King George V declared the Mersey Tunnel open to traffic – the same Tunnel as we are using today.

I am confident that this new complex of roads, viaducts and marshalling areas will improve the flow of traffic through the tunnel and, of no less importance to Birkenhead, will help to relieve congestion in the centre of town for many years to come.

The initial entrance point for this flyover is from the A553 Conway Street and the A552 Borough Road. The Borough Road Flyover is still in existence today, but the Conway Street Flyover was demolished in the early 1990s as part of the Wirral Citylands Redevelopment within Birkenhead town centre. It was decided that the Conway Street Flyover would be demolished and a roundabout along Argyle Street would be provided for the purpose.

Queensway Tunnel entrance from Birkenhead, *c.* 1970. Note the new booths and Marshalling Yard markings. (Wirral Archives)

Birkenhead Entrance – somewhat less attractive today due to the 1969 changes to the Marshalling Yard. (Author)

BPL 0240. Conway Street before the flyover had been constructed. (Birkenhead Library)

BPL 3294. (Birkenhead Library)

BPL 3295.
(Birkenhead Library)

BPL 3307.
(Birkenhead Library)

BPL 3309.
(Birkenhead Library)

Completed Chester Street approach. (Birkenhead Library)

A41 approach to tunnel and underpass. (Author)

Remainder of Old Flyover to Conway Street, Birkenhead. (Author)

View down underpass to approaching traffic with Chester Street exit to the right-hand side of the picture. (Author)

The following 'Birkenhead Entrance to the Queensway Tunnel Flyover Construction and Demolition' pictures were kindly provided by the Birkenhead Central Library:-

BPL 0940 aerial view of 1969 tunnel access works.

Showing extent of clearance areas for the new tunnel access roads.

Construction of flyover and amount of steel reinforcement within it for the proposed heavy traffic use.

General site photograph showing main drainage sections waiting being fitted.

Partly demolished flyover. Directly between the sections the main road – close to the centre of Birkenhead – had to be closed at week-end for the removal and demolition works.

Two independent supports awaiting final demolition.

Section in front of the former Conway School – now Council One Stop Shop – on the edge of the town centre, awaiting final demolition.

Heavy lifting equipment at weekends to remove roadway sections safely and with minimal traffic disturbance.

Sections during demolition. Note the steel reinforcement being separated from the concrete during this phase.

Emergency Exits

Concerns were raised about how to improve the capacity for emergency evacuation in case of an incident in one of the Mersey Tunnels. The original configuration of the tunnel-links between the tunnels was not entirely suitable as an escape provision. Discussions between Mersey Tunnels and Merseyside Fire Brigade identified the need for additional links between the road tunnels to satisfy their evacuation strategy in the case of an incident within one of the road tunnels. It was considered that the provision of three additional cross passages within the central section of the tunnels, located between the two existing ventilation shafts, satisfied the objectives of the evacuation strategy.

Mersey Tunnels commissioned Mott MacDonald to investigate and design the passages in the Kingsway, as well as the emergency refuges in the Queensway Tunnel. It was decided that seven emergency refuges – named from A to G – were to be used in the Queensway Tunnel and that three routes – nicknamed Tom, Dick and Harry by Mersey Tunnels – were to be used in the Kingsway Tunnel. These were positioned equally along the tunnel to miss the geological faults and to be located between the existing ventilation shafts. This decision placed the new passages at 325 metre intervals and the decision was made that, given the tunnel's location and the other aspects of the design, the scheme could be given the green light.

In April 2004, work started on each of the seven new Queensway Tunnel refuges. The refuges had a capacity of 180 people and was the start of a £9 million project to bring the tunnels in line with the highest European Safety Standards. Each refuge is 21 metres long (69 feet) and 3 metres wide (9.8 feet), and is accessible from the main tunnel walls. The refuges have fire resistant walls, ramps for wheelchair access, a good supply of bottled water, toilet facilities, and a direct video link to the Mersey Tunnel Police Control Room. All refuges are linked by a walkway below the road surface with exits at both the Liverpool and Birkenhead exists of the tunnel.

The existing shafts were to be used for new exits and the Georges Dock exit was to utilise the second shaft, which had been sunk on the site of Georges Dock, to facilitate the task of excavation. At Morpeth Dock, which had only one shaft on this side of the river, it was decided that this should be utilised for the same purpose.

Passengers would exit from the tunnel to the shafts via a newly constructed staircase. Originally the entrances to this escape routes were distinguishable from a considerable distance away, depending where you were in the tunnel at the time. The use of an illuminated sign with the title 'Emergency Exit' and a red arrow pointing in the direction of the exit were the only clues you had, and in thick black smoke these would not have proved efficient enough. On the opposite wall of the tunnel a similar illuminated sign was erected to ensure the exits were clear.

The Georges Dock exit leads along a passage and a short final staircase to the roadway close to the ventilation station. The Morpeth Dock exit is situated in Shore Road, Birkenhead, close to the old Canning Street railway building and connected by a passage and staircase leading from the shaft. The passage from the Morpeth Dock shaft to the final staircase had to be constructed in circular form, with a cast-iron lining, to protect it from water in the surrounding strata. Otherwise, all these passages and staircases were constructed in the ordinary way, and were lined with concrete.

Work on the refuges' full scale mock-up designs started in May 2003 to determine the design and how to meet the optimum design requirements. Design of the cross passages focussed on areas that would determine any spatial requirements within the confines of the existing tunnels for wheelchair and mobility-impaired access, as well as passenger escape requirements. With this in mind, the contractor would then be able to ascertain which construction method to use.

At the Birkenhead end of the tunnel (Shore Road) a 45-metre-long, 3-metre-diameter tunnel was bored through sandstone to provide additional access to the existing emergency staircase leading to the surface at Shore Road. Traditionally, hand excavation techniques involving miners using handheld compressed air tools would have been employed.

British tunnels have so far had a good safety record through a combination of good management, the installation of safety systems to the latest European standards, and an information programme to make people aware of what they should do in a tunnel emergency. The actual levels and positions of the two Mersey Tunnels were surveyed as a part of a comprehensive investigation before commencement of the new refuge and other works.

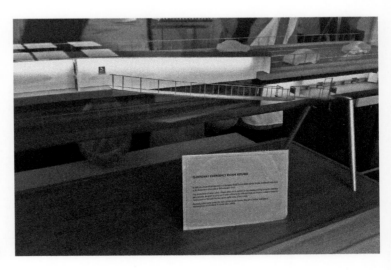

Model of typical
escape refuge. (Author)

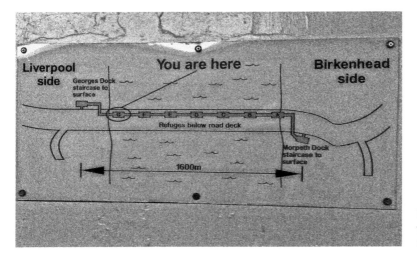

Location sign within the tunnel refuge. (Author)

Reassuring sign within tunnel refuge. There are also audio and video links to the Emergency Services. (Author)

Shore Road Tunnel emergency exit. (Author)

Old Equipment in Liverpool Museum Storage Today and Memorable Items

The following images were taken by the author during the research for this book.

An early example of a tunnel cleaning truck.

A later example of a tunnel cleaning truck.

Cleaning boom and scrubber pad.

Tunnel scrubber pad and end of boom.

Internals of the early version of the tunnel maintenance wagon. Note the sparse features with the seat removed.

Could you sit next to that engine whilst travelling each way along the tunnel length?

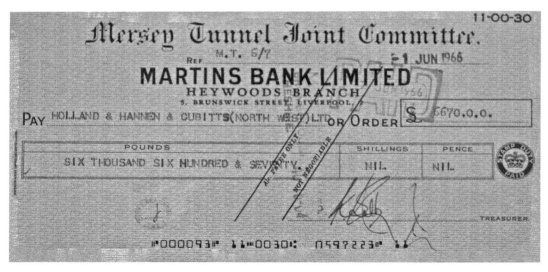

1966 cheque to Cubbits, one of the many cheques issued during the construction of the Queensway Tunnel.

Above left: Original Tunnel commemorative coin.

Above right: Commemorative Tunnel coin.

Kingsway 'New Tunnel Walkthrough' ticket, 27 July 1971.

The black basalt figures signify day and night showing the tunnel is always open.

Tunnellers' Memorials

During blasting for the tunnels and ventilation shafts the workers were warned to retreat to particular point that was deemed safe. Unfortunately, what the engineers were not aware of was a large void above the area. This void was supposed to have been created by attempts to undermine the outer wall of Liverpool castle during a conflict. This area collapsed as a result of the blast, sadly dropping tonnes of rock onto some of the men in the tunnel.

As a result of this tragedy, and perhaps due to the young age of the victim in question, a labourer named James Herbert Brown, aged just eighteen, was killed. His colleagues requested that they be allowed to attend his funeral but the directors refused, but as a mark of respect the whole workforce took the day off without pay and attended his funeral. It has been said that it was a sight to behold as scores of men escorted his coffin down Smithdown Road to Toxteth cemetery. It was James's relative – also called James after him – who started to cause a fuss about the lack of a memorial and was victorious when a plaque naming all those who lost their lives was erected on the ventilation shaft at Mann Island during the tunnel's jubilee celebrations.

There is a plaque to each of the tunnels – shown below – detailing those that paid the price during the construction of the tunnels. The Queensway plaque can be found on the Georges Dock building, and the Wallasey (Kingsway) plaque can be found along the Wallasey promenade ventilation shaft, close the Seacombe ferry terminal.

An example of one of the brave men who constructed the Queensway Tunnel is Mr Joseph McNulty, who lived at 337 Price Street, Birkenhead. At his inquest, the coroner – Mr Joseph Roberts – along with a jury heard the account of Mr McNulty's accident and subsequent injuries. Mr John Blakely, a miner living at 8 Planet Street, Rock Ferry, said that at 5.45 a.m., he and six others were working on scaffolding near the mid-tunnel. Staging was 18 feet high (5.486 metres) and suspended on bearers and planks. Mr McNulty was going to put two bolts in a plate on the side of the tunnel and stood on a rock ledge and the side of the tunnel to move a plank. The plank was wet and greasy and Mr McNulty slipped trying to maintain his hold and fell to the floor. Witnesses went to his aid and Mr McNulty was taken to hospital. Witnesses said the accident happened 2 to 3 minutes before the end of the duty. Mr McNulty was admitted to the general hospital at 7 a.m. with a broken right arm, cuts to his face – which turned septic – and bronchial pneumonia, which resulted in the death of Mr McNulty three days later.

Queensway Tunnel (Birkenhead) Memorial

Memorial stone of St George's Dock building to Queensway tunnel workers who died during its construction Stone reads: THIS MEMORIAL WAS ERECTED AS PART OF THE QUEENSWAY TUNNEL DIAMOND JUBILEE CELEBRATIONS IN REMEMBRANCE OF THOSE WHO DIED DURING ITS CONSTRUCTION. WORK ON THE TUNNEL WAS COMMENCED ON 16TH DECEMBER 1925 AT THIS SITE BY HER ROYAL HIGHNESS PRINCESS MARY AND THIS GREAT FEAT OF ENGINEERING WAS OPENED AT OLD HAYMARKET ON 18TH JULY 1934 BY HIS MAJESTY KING GEORGE V. (Author)

The following are commemorated upon the memorial:-

JOSEPH BONNER
26th July 1926 Aged 26

JOHN JOSEPH McNULTY
17th September 1928 Aged 29

JAMES HERBERT BROWN
29th November 1928 Aged 18

JOHN WILLIAM BLAKELEY
7th July 1929 Aged 34

JOHN McNICHOLAS
13th September 1929 Aged 55

HENRY FRANCIS GARRETT DE MOUL
28th September 1929 Aged 25

JOSEPH COLLEGE
11th December 1929 Aged 62

JAMES MICHAEL WILMOTT
27th December 1929 Aged 42

JOHN CARBERRY
24th March 1930 Aged 26

ALBERT WHITE
27th November 1930 Aged 42

ALFRED PITMAN DUKE
16th July 1931 Aged 45

HENRY DETITH
15th September 1931 Aged 24

FREDERICK JOSEPH DURR
29th September 1931 Aged 33

THOMAS ARTHUR BECKINGHAM
16th May 1933 Aged 57

JOHN CARR
14th July 1933 Aged 23

JAMES GREEN (Bir/182/53)
14th November 1933 Aged 22

DONALD LESTER
15th September 1934 Aged 24

Kingsway Tunnel (Wallasey) Memorial

Memorial Stone on Wallasey ventilation station regarding Kingsway tunnel workers who died during its construction Inscription at bottom reads: THIS MEMORIAL WAS ERECTED IN 1997 IN REMEMBRANCE OF THOSE WHO DIED DURING THE CONSTRUCTION OF THE KINGSWAY TUNNEL BETWEEN 1966 AND 1974. (Author)

The following are commemorated upon the memorial:-

FRANCES ADDERLEY
15th June 1970 Aged 59

RONALD (ROY) CARRY
12th June 1969 Aged 26

BERNARD GLENN DENNESS
1st November 1969 Aged 25

CHARLES KEEGIN
1st November 1969 Aged 28

JOHN LATHAM
23rd October 1971 Aged 33

LEONARD MILLS
8th March 1969 Aged 47

JOZSEF NYARI
23rd October 1971 Aged 32

GERALD RANDELS
3rd March 1971 Aged 38

DANIEL JOSEPH SWEENEY
1st November 1969 Aged 27

Bibliography

I would like to thank the following organisations and groups that have assisted me in the research for this book by allowing me access to their numerous records:

Archives kept in Liverpool Central Library.

Wonders of World Engineering magazine at http://www.wow-engineering.co.uk/. This site contains magazine articles from a magazine published in 1937. They are a detailed historical view of various engineering feats. The site includes a copy of an article on the construction of the Mersey Tunnel from Birkenhead to Liverpool.

Two booklets published by the old Mersey Tunnel Joints Committee. They are called 'Mersey Tunnels – The Story of an Undertaking' and 'Mersey Tunnels 2'.

Wirral Archives, for their numerous and vast document availability.

Merseyside Museum Large Objects Store and documents library.

Liverpool Library, for their numerous documents and Microfiche items.

'The Mersey Tunnel. The Story of an Undertaking' – Merseytravel

'Story of the Mersey Tunnels (Officially Named Queensway)' – Merseytravel

'All about the New Mersey Tunnel' (Pamphlet) – Merseytravel

'The Mersey Tunnel (Queensway)' (Pamphlet) – Merseytravel

The Birkenhead Mersey Tunnel Approach Scheme July 1969.

Stanton Iron Road and the Mersey Tunnel. (Liverpool Museum)